ers of Bernoulli. See Note G. (page 722 *et seq.*)

			Working Variables.					B_1 dec. fr.	B_3 dec. fr.	B_5 dec. fr.	
$^0V_8$ 0 0 0	$^0V_9$ 0 0 0	$^0V_{10}$ 0 0 0	$^0V_{11}$ 0 0 0								$^0V_{24}...$ 0 0 0
								B_1	B_3	B_5	B_7
...	...	...	$\dfrac{2n-1}{2n+1}$								
...	...	...	$\dfrac{1}{2} \cdot \dfrac{2n-1}{2n+1}$								
...	...	...	0		....	$-\dfrac{1}{2} \cdot \dfrac{2n-1}{2n+1} = A_0$					
...	...	$n-1$									
...	...	...	$\dfrac{2n}{2} = A_1$								
...	...	...	$\dfrac{2n}{2} = A_1$	$B_1 \cdot \dfrac{2n}{2} = B_1 A_1$	....			B_1			
...	...	...	....	0	$\left\{-\dfrac{1}{2} \cdot \dfrac{2n-1}{2n+1} + B_1 \cdot \dfrac{2n}{2}\right\}$						
...	...	$n-2$									
$\dfrac{n-1}{3}$											
0	...	...	$\dfrac{2n}{2} \cdot \dfrac{2n-1}{3}$								
...	$\dfrac{2n-2}{4}$	...	$\left\{\dfrac{2n}{2} \cdot \dfrac{2n-1}{3} \cdot \dfrac{2n-2}{3} = A_3\right\}$								
...	0	...									
...	...	...	0	$B_3 A_3$	....				......	B_3	
...	...	...	....	0	$\left\{A_3 + B_1 A_1 + B_3 A_3\right\}$						
...	...	$n-3$									

ons thirteen to twenty-three.

| ... | ... | ... | .... | .... | .... | | | ...... | | | B_7 |

ADA LOVELACE

ADA LOVELACE
THE MAKING OF A COMPUTER SCIENTIST

Christopher Hollings,
Ursula Martin & Adrian Rice

Bodleian Library
UNIVERSITY OF OXFORD

First published in 2018 by the Bodleian Library
Broad Street, Oxford OX1 3BG
Reprinted 2019

www.bodleianshop.co.uk

ISBN 978 1 85124 488 1

Text © the contributors, 2018
All images, unless specified,
© Bodleian Library, University of Oxford, 2018

All rights reserved.

This book is the result of a collaboration with the Clay Mathematics Institute.

 Clay Mathematics Institute

The publisher gratefully thanks Lord Lytton for his kind permission to reproduce material in this book from the Lovelace–Byron archive on deposit at the Bodleian Library.

No part of this book may be reproduced, stored in a retrieval system, or transmitted in any form or by any means, electronic, mechanical, photocopying, recording, or otherwise, without the written permission of the Bodleian Library, except for the purpose of research or private study, or criticism or review.

Cover design by Dot Little at the Bodleian Library
Designed and typeset in 11 on 14 Monotype Bulmer
by illuminati, Grosmont
Printed and bound in China by 1010 Printing International Ltd.
on 140 gsm Chen Ming Bi Yun Tian paper

British Library Catalogue in Publishing Data
A CIP record of this publication is available from the British Library

CONTENTS

	PREFACE	vii
	ACKNOWLEDGEMENTS	ix
	DRAMATIS PERSONAE	x
1	Ada Lovelace and the advancement of science	1
2	A mathematical childhood	9
3	Early influences	23
4	Charles Babbage and the thinking machine	37
5	Learning mathematics with Professor De Morgan	49
6	Becoming a mathematician	61
7	The Analytical Engine	73
8	Mathematical puzzles and speculations	89
9	The final years	99
	NOTES	105
	FURTHER READING	109
	IMAGE CREDITS	110
	INDEX	112

PREFACE

Ada, Countess of Lovelace, is often called the 'first computer programmer'. Her 'first program' – a large table with mathematical entries – appeared in 1843, in an article about the Analytical Engine. This remarkable mechanical computer was designed by the mathematician and inventor Charles Babbage, but never built. Lovelace's article is a translation from the French of a shorter paper by Luigi Menabrea, with substantial appendices written by Lovelace which are nearly twice the length of the original work. It contains an account of the principles of the machine and far-sighted ideas of what it might be able to do, which still capture the imagination today.

In this book we draw on the archives of family correspondence to reveal Ada Lovelace's unusual and eclectic education. Born in 1815, Ada was the daughter of the Romantic poet Lord Byron and his highly educated wife, Anne Isabella. Her parents separated when she was one month old, and she was brought up by her mother. She was an imaginative and intelligent child, and developed her interests in mathematics and science from an early age. In an era when university and much school education was closed to women, she studied with governesses and private tutors and got to know the scientific elite of the day, including

1 Ada Lovelace, or The Right Honorable Lady King as she was then known, aged twenty in this 1836 portrait by Margaret Carpenter. Ada wrote at the time: 'I conclude she is bent on displaying the whole expanse of my capacious jaw bone, upon which I think the word Mathematics should be written.'

Mary Somerville and Charles Babbage. After her marriage, and the births of three children, she studied advanced mathematics. Over an intense eighteen-month period, an exchange of around sixty letters with the noted mathematician Augustus De Morgan shows her learning university-level calculus and developing her mathematical strengths: attention to detail, an ability to think from first principles, and an imaginative approach to big questions. The archive also sheds light on Ada Lovelace's work with Charles Babbage, her writings on his Analytical Engine, and how she maintained a variety of scientific and social interests until her tragically early death at the age of thirty-six.

The Lovelace–Byron archive, a deposited collection of some 400 boxes of family papers, is held in the Bodleian Library of the University of Oxford. In 2016, the Clay Mathematics Institute (CMI), in partnership with the Library and with the strong support of the Earl of Lytton, launched a project to make the mathematical papers in the archive more widely available. The first phase involved digitization of her correspondence with Augustus De Morgan: images of their letters were posted on CMI's website (www.claymath.org) alongside transcripts and commentary by Christopher Hollings. This book takes the project forward. It makes images of some of the papers from the archive available for the first time.

ACKNOWLEDGEMENTS

The authors acknowledge the generous support of the Bodleian Libraries, the Clay Mathematics Institute and the UK Engineering and Physical Sciences Research Council. We are particularly grateful to June Barrow-Green, Geoffrey Bond, Mary Clapinson, Richard Holmes, Adrian Johnstone, Ann Kettle, Peter Neumann, Sydney Padua, Murray Pittock, Miranda Seymour, Doron Swade, Sophie Waring, Robin Wilson and Stephen Wolfram for helpful discussions.

DRAMATIS PERSONAE

Albert of Saxe-Coburg and Gotha, Prince Consort (1819–1861)   Husband of Queen Victoria. A strong supporter of the sciences and promoter of the Great Exhibition of 1851.

Anne Isabella ('Annabella') Noel Byron (née Milbanke), 11th Baroness Wentworth and Baroness Byron (1792–1860)   Educational reformer and wife of Lord Byron.

Charles Babbage (1791–1871)   Mathematician, scientist and inventor of the Analytical Engine.

Isambard Kingdom Brunel (1806–1859)   Engineer whose railways, tunnels and bridges revolutionized public transport and civil engineering.

George Gordon Byron, 6th Baron Byron (1788–1824)   Lord Byron, the influential poet and leading figure in the Romantic movement.

Augustin-Louis Cauchy (1789–1857)   French mathematician regarded as a pioneer of the mathematical area of analysis, which gave a rigorous logical basis for calculus.

George Cayley (1773–1857)   Pioneer of aeronautical engineering, who discovered the mathematical principles of flight and built the first piloted glider.

Robert Chambers (1802–1871)   Scottish publisher, geologist

and anonymous author of the controversial book *Vestiges of the Natural History of Creation* (1844).

JOSEPH CLEMENT (1779–1844)   Engineer and instrument maker, who worked on Charles Babbage's first Difference Engine between 1824 and 1832.

CHARLES DARWIN (1809–1882)   Biologist and geologist, best known for his theory of natural selection in the study of evolution.

AUGUSTUS DE MORGAN (1806–1871)   Mathematician, logician and first Professor of Mathematics at University College London.

SOPHIA DE MORGAN (née FREND) (1809–1892)   Social activist, daughter of William Frend and wife of Augustus De Morgan.

CHARLES DICKENS (1812–1870)   Writer known personally to both Lovelace and Babbage. Regarded as one of the greatest novelists of the Victorian era.

LEONHARD EULER (1707–1783)   Swiss mathematical scientist who made numerous important discoveries in pure and applied mathematics.

MICHAEL FARADAY (1791–1867)   Scientist who made major contributions to electromagnetism, electrochemistry and the popularization of science.

WILLIAM FREND (1757–1841)   Mathematician, nonconformist social reformer and writer.

AGÉNOR DE GASPARIN (1810–1871)   French statesman and writer on agricultural matters.

WORONZOW GREIG (1805–1865)   Eldest son of Mary Somerville by her first marriage, a friend and correspondent of Ada Lovelace.

WILLIAM ROWAN HAMILTON (1805–1865)   Irish physicist, astronomer and mathematician.

CAROLINE HERSCHEL (1750–1848)   Astronomer who discovered several comets. The sister and collaborator of William Herschel (1738–1822), the discoverer of Uranus.

She and Mary Somerville were the first female fellows of the Royal Astronomical Society.

JOSEPH MARIE JACQUARD (1752–1834) French merchant and inventor of the 'Jacquard loom', which used punched cards, inspiring their use for programmable machines, such as Babbage's Analytical Engine.

AUGUSTA ADA KING (née BYRON), COUNTESS OF LOVELACE (1815–1852) Mathematical writer, often described as the first computer programmer.

WILLIAM KING (1786–1865) Physician, philanthropist and friend of Lady Byron.

WILLIAM KING, 1ST EARL OF LOVELACE (1805–1893) Landowner with an interest in agricultural economics, who married Ada Byron in 1835.

DIONYSIUS LARDNER (1793–1859) Irish physicist and scientific writer who worked mainly as a popularizer of science, and wrote a popular account of Babbage's difference engine.

HARRIET MARTINEAU (1802–1876) Social theorist and author of a variety of texts on sociological, religious and economic matters, often from a feminist perspective.

KARL MARX (1818–1883) German political philosopher, economist and social scientist, remembered today as one of the founders of communism.

LUIGI MENABREA (1809–1896) Italian military engineer, who later served as the prime minister of Italy. His account of Babbage's Analytical Engine formed the basis of Ada Lovelace's key work.

FLORENCE NIGHTINGALE (1820–1910) Social reformer and statistician, regarded as the founder of modern nursing.

LOUIS POINSOT (1777–1859) French mathematician and physicist, best known for his work in geometry.

ADOLPHE QUETELET (1796–1874) Belgian astronomer, mathematician and statistician, who introduced statistical methods to the social sciences.

Mary Somerville (née Fairfax) (1780–1872) Popular scientific author and polymath.
Alan Turing (1912–1954) Pioneering mathematician, logician and cryptanalyst, influential in the development of theoretical computer science and artificial intelligence.
Charles Wheatstone (1802–1875) Scientist, inventor and major figure in the development of telegraphy. Served as editor of Richard Taylor's *Scientific Memoirs*.

AUTHORS' NOTE

The Honourable Ada Byron, sometimes known as Miss Byron, became the Lady King on her marriage in 1835, and the Countess of Lovelace, sometimes known as Lady Lovelace, when her husband became the Earl of Lovelace in 1838. For clarity we use Ada, Ada Byron or Ada Lovelace in the text.

1
ADA LOVELACE AND THE ADVANCEMENT OF SCIENCE

Augusta Ada Byron was born on 10 December 1815 at the central London family home of her father George Gordon, Lord Byron, the most famous poet in England at the time. A month after Ada's birth her parents separated, and her father moved to France. He lamented the parting in his widely read long poem *Childe Harold's Pilgrimage*:

> Is thy face like thy mother's, my fair child!
> Ada! sole daughter of my house and heart?
> When last I saw thy young blue eyes, they smiled,
> And then we parted, – not as now we part,
> But with a hope.[1]

Byron never saw his daughter again. He died of a fever in 1824 while fighting in the Greek War of Independence.

Ada was brought up by her mother, Anne Isabella, known as Annabella. She was the daughter of Sir Ralph Milbanke, a Member of Parliament with substantial estates and mining interests in County Durham, and one of an influential group of intellectuals and social reformers based in the North of England. The future Lady Byron received a good education for the time, not untypical for a woman of her background, studying with private tutors. She

2 Lord Byron in 1813, painted by Richard Westall.

learned mathematics and astronomy from William Frend, a radical thinker, well known for both his nonconformist religion and his odd mathematical ideas. Byron, whose mathematical education was more modest and who had great difficulty even keeping his accounts straight, initially praised his wife as the 'Princess of Parallelograms',[2] but subsequently parodied her, in his famous poem *Don Juan*, as Donna Inez, whose 'favourite science was the mathematical'[3] and whose 'thoughts were theorems'.[4] In later life, Lady Byron was recognized as an educational reformer, and established a number of schools which followed the principles of the Swiss educational reformer Pestalozzi, in combining book learning with physical work and practical skills. She is one of sixty-three people, including John Stuart Mill, Joseph Priestley and Elizabeth Fry, commemorated on the Reformers' Monument in London's Kensal Green Cemetery.

3 Sketch of Lord and Lady Byron, about 1815, by his former lover Lady Caroline Lamb, who called him 'mad, bad and dangerous to know'.

Ada Lovelace's life spanned the growth of extraordinary prosperity in Britain, from the end of the debilitating Napoleonic Wars in 1815 to the triumphant Great Exhibition of 1851. This was visited by over six million people, who viewed displays from twenty-five countries, with pride of place given to Britain's engineering and technical achievements. Key to this innovation was steam power. Richard Trevithick's 1801 'Puffing Devil' was the first steam-driven vehicle; the first commercial railway opened in 1825; and new steam-powered machinery, and other engineering innovations, transformed manufacturing. Britain's stable politics and years of peace encouraged investment in railways, factories and shipping, which enabled the exploitation of the resources of Britain and the Empire, and led to unprecedented economic growth.

Not just the scientific elite, but a growing number of non-scientists, female and male, were becoming interested in the latest ideas. Public lectures and demonstrations were popular, and the annual meeting of the newly founded British Association for the Advancement of Science attracted several hundred people. Mathematics was an increasing part of these developments, with a growing understanding of its importance in the study of natural and social phenomena – observing the stars, recording the tides, or analysing the harvests. Making use of these results for navigation, engineering or agriculture increased demand for reliable

4 Lady Byron, Anne Isabella Milbanke, in 1833, engraving by William Henry Mote after a portrait by William John Newton.

5 The main hall of the Great Exhibition from *Dickinson's Comprehensive Pictures of the Great Exhibition*, 1851.

information, in the form of published tables, all produced by hand calculation. The science of statistics was emerging, with Florence Nightingale's analysis of data from the Crimean War providing powerful evidence for the need to fight infection in hospitals (FIGURE 6). Although doing new mathematical research remained largely an activity for the gentleman amateur or those employed to do other things – such as teaching in universities, serving in the army and navy or working as actuaries – opportunities for educated men, and a few women, to engage with the mathematical sciences were increasing.

During the nineteenth century, philanthropic initiatives, such as Lady Byron's schools, saw free elementary education becoming available for nearly everyone. Mathematics took the form of 'practical mathematics', the arithmetic and geometry needed for accounting, surveying or navigation. Those with the opportunity to learn more, at schools for boys, and more informally for girls, might, like Lady Byron herself, have learned enough algebra to solve simple equations, or have studied geometry while making astronomical observations. Euclidean geometry was widely studied and seen as valuable for developing habits of rigorous thought, as well as for its practical use. It starts with a small set of statements about lines and points, and uses these according to certain fixed rules to form proofs of theorems about geometrical shapes. Until the early 1900s most university students had to study Euclid too, often passing examinations by memorizing the complicated proofs. Lord Byron had attended Cambridge University, but did not have to take any such examinations, as members of the House of Lords were exempt.

Mathematics was surprisingly widespread in popular culture: young ladies did examples from Euclid for pleasure; periodicals like *The Ladies' Diary* published mathematical questions and readers' answers; astronomy was a popular pursuit; and poets

6 Florence Nightingale's 'Rose Diagram', demonstrating that the primary cause of death among British forces in the Crimean War was infectious disease (blue areas), not injury (red areas). It forms the frontispiece of the campaigning book *England and Her Soldiers*, written by Nightingale and Harriet Martineau in 1859.

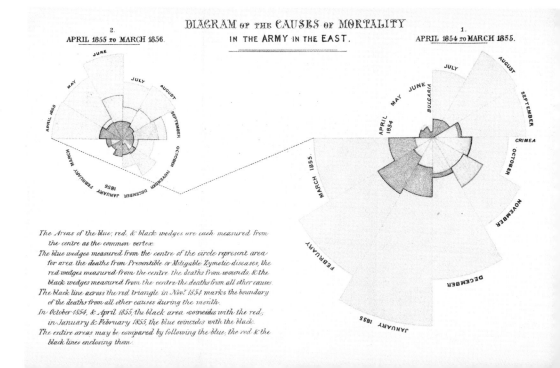

sometimes used complex mathematical imagery. Coleridge, for example, described how a flock of starlings 'shaped itself into a circular area, inclined – now they formed a Square – now a Globe – now from complete orb into an Ellipse ... now a concave Semicircle'.[5]

More advanced studies, in universities or military academies, might include the calculus and geometry needed to understand interest calculations, tides or gunnery. Ada Lovelace's early teachers, William Frend and Dr William King, had had this kind of education: her later mentors, Charles Babbage and Augustus De Morgan, were at the heart of attempts to change it. They wanted to modernize mathematical education to reflect new ideas

7 Lady Byron in the audience at the First World Anti-Slavery Convention in 1840, organized by the Quaker William Sturge. In this 1841 painting by Benjamin Robert Haydon she appears to the left of the woman in the large black hat.

8 Images of student life at Somerville, one of the first colleges for women in Oxford, from *The Graphic*, July 1880.

emerging from France and Germany, in which calculus, like Euclidean geometry, was subject to rigorous deduction from rules.

In the first half of the nineteenth century, talented women like Mary Somerville or Ada Lovelace could only learn more advanced mathematics through private study and informal access to libraries or scientific meetings. But the situation slowly began to improve. Augustus De Morgan taught at the Ladies' College, at Bedford Square in London, founded in 1849 and one of the first institutions in the world to offer college education to women. Girton College, the first college open to women in Oxford or Cambridge, was founded in 1869, and several hundred women had studied mathematics to degree level by the turn of the century.

2
A MATHEMATICAL CHILDHOOD

Lady Byron's 'happy and intelligent'[1] daughter started to learn her letters from her mother when she was nearly three. A Miss Lamont was appointed as a temporary governess when Ada was five. Lady Byron imposed a demanding schedule:

> lessons in the morning in arithmetic, grammar, spelling, reading, music, each no more than a quarter of an hour long – after dinner, geography, drawing, French, music, reading, all performed with alacrity and docility.[2]

The discipline seems harsh – though not unusual for the time. Ada was required to lie still on a board, and was punished if she fidgeted. A system of tickets was used for reward and punishment, with any surplus spent on an improving book.

Miss Lamont was required to keep a diary for the often absent Lady Byron. In May 1821 she wrote:

> Commenced giving instructions to Miss Byron ... The first trial was in arithmetic. She adds up sums of five or six rows of figures, with accuracy; she is deliberate and correct in the process, and takes an interest in the performance.[3]

9 Ada Byron aged six, painted for her father by the society painter Count Alfred d'Orsay, c. 1821.

10 *The Governess*, painting by Richard Redgrave, 1844.

The governess was also required to write letters from Ada to Lady Byron:

> I want to please Mama very much, that she and I may be happy together. ... The French has not interested so much as some others – and one night I was rather foolish in saying that I did not like arithmetic & to learn figures, when I did I was not thinking quite what I was about. The sums can be done better, if I tried, than they are. The lying down might be done better, I might lay quite still never move.[4]

There were a few diversions – learning to sew, walks in the garden and to 'play at being a horse, during which the neighing of the horse might be heard to a considerable distance'.[5]

Pestalozzi's principles encouraged learning through practical skills:

The morning employed in building with the wooden bricks of the brick-box an attempt at forming some little designs from Pestalozzi – Ada became highly animated in the occupation, and particularly delighted with the idea of enclosing a city and raising towers within the walls – but took more pleasure in imagining for herself as she proceeded, than being guided by the model.[6]

11 Ada Byron's daily activities, aged six, which included music, French, arithmetic, needlework and exercise out of doors.

Ada describes dissecting a dragon-fly:

> We opened the head and the eye of a dragonfly – and we opened its mouth, & we saw its tongue, it was a very little pink thing – It was caught and put under a glass when alive – and when we opened its eyes and saw the things, it was dead then.[7]

She was fascinated by geography, which amuses me very much – there are wild forests in Norway & the sea-coasts are very dangerous & the waves are higher than this house sometimes – In Iceland there are curious sorts of mountains that make curious sorts of noise – & there is fire that comes out – & a liquid runs down the mountain.[8]

12 Like many at the time, the young Ada Byron was fascinated by reports of volcanoes. Her diary refers to Hekla in Iceland, which erupted in 1768. In 1811 a volcanic eruption under the sea created Sabrina Island, off the Azores, as graphically described in the 1821 children's book *Wonders!*, published by Harris & Son.

Well! this is a wonder of wonders to me!
Such volumes of fire bursting out from the sea!
With lava, and ashes, and sulphurous smell,
I'm surpris'd that the sailors can bear it so well!
Yet all must desire the eruption to view,
And if I were there I might feel anxious too.

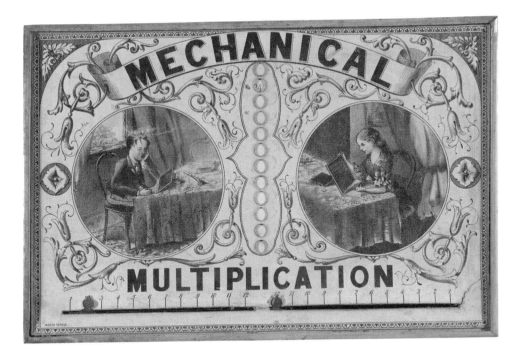

As Ada grew older, her mother seems to have taken on more of the teaching. Ada's letters to her bubble over with enthusiasm – Italian, music, drawing, geography, her cat and unexpected kittens, summaries of the Sunday sermon, and her mathematics. Aged ten, she wrote about the so-called rule of three:

> I have been puzzling hard at a sum in the rule of three which I could not do, the question is 'If 750 men are allowed 22500 rations of bread per month, how many rations will a garrison of 1200 men require'?[9]

The rule of three is a procedure to find an unknown quantity, z, given three known numbers, a, b and c. It says that, if

$$\frac{a}{b} = \frac{c}{z}$$

then

13 Mechanical Multiplication: a nineteenth-century mathematical puzzle for children.

A MATHEMATICAL CHILDHOOD

"How many different kinds of earth have you observed?
6. viz: chalk, gravel, clay, mould, sand and sla[te]
"What are the uses of each kind in agriculture, art[s]
and manufactures?"
Chalk is used in medicine, for building, for mor[tar]
for making starch, and for marking with.—
Gravel is used for making walks and roads—
Clay is used for making china, for bricks and
tiles, and for forming the bottoms of canals and
ponds.— Mould is used for growing plants upon.
Sand is used for making glass, for mortar, for
earthenware, for cleaning pots and pans, for
manure and for drying writing.— Slate is used for
roofing houses, for writing upon, and for drawing &
writing with; for cleaning iron, for making fences,
and for ornamental vases & inkstands.—
"Were the first settlements of men, made on mounta[ins]
or in vallies?"
The vallies.
"Why is one situation superior to the other?"

situation in a valley is superior to one on a
mountain, because it is much more sheltered & accessible,
valleys are also much more fertile than mountains.

I will answer the other questions about the valleys &
mountains, as soon as Mama has got the books
she mentioned.

I had not till Mama showed me, understood
the Sums ~~that were in~~ where both multiplication and
addition signs were used, in consequence of which
the former examples must have been wrong, as
I thought they were as follows: $5 \times 2 = 10$ to which
add $3 = 13$ ~~which~~ multiplied by 4 gives 52, instead of
$5 \times 2 = 10$ to which add $3 \times 4 = 12$

$2 \times 5 = 10$
$2 \times 5 \times 3 = 30$
$3 \times 8 \times 7 \times 4 = 672$
$4 \times 5 \times 6 \times 7 = 840$
$5 \times 6 \times 7 \times 8 = 1680$
$6 \times 7 \times 8 \times 9 = 3024$
$7 \times 8 \times 9 = 504$
$8 \times 9 \times 11 = 792$
$9 \times 11 \times 12 = 1188$
$11 \times 12 \times 13 = 1716$

$2 \times 5 + 3 \times 4 = 22$
$3 \times 4 + 5 \times 6 = 42$
$4 \times 6 + 5 \times 7 = 59$
$5 \times 6 + 7 \times 8 = 86$
$6 \times 7 + 8 \times 9 = 114$
$6 \times 8 + 7 \times 9 = 111$
$6 \times 9 + 7 \times 8 = 110$
$7 \times 11 + 12 \times 13 = 233$
$7 \times 12 + 11 \times 13 = 227$

$$z = b\frac{c}{a}$$

In Ada's problem, a is 750, b is 22,500 and c is 1,200. Therefore, by simple multiplication and division, the answer should be

$$z = 22{,}500 \times \frac{1{,}200}{750} = 36{,}000$$

In the same letter, Ada said she was 'very desirous to master those overs in the sums of division'[10] and added

> I think by the time you come back I may have learnt something about decimals, I attempted the double rule of three but I could not understand, however I will not give it up yet.[11]

The double rule of three is similar to the rule of three, but involves more complicated fractions. Pupils were usually expected to remember just the formula, but Ada wanted to understand. As she explained to her mother, 'the book does not teach as well as you do'.[12]

Ada was introduced to Euclid at the age of twelve, thanking her mother for the 'entertaining pamphlet you left me on that subject..., which I found very amusing indeed'. Although confessing that 'I am a little afraid of the Theorems', she resolved to 'attack them boldly & do my best'.[13]

An exercise sheet written in February 1829 (FIGURE 14), when she was just thirteen, contains answers to numerous geography questions about rivers, mountains and valleys, before turning to arithmetic. We see Ada wrestling with the problem of working out $5 \times 2 + 3 \times 4$:

> I had not till Mama showed me, understood the sums where both multiplication and addition signs were used, in consequence of which the former examples must have been wrong, as I thought they were as follows:

PREVIOUS PAGES
14 Ada Byron reflecting on the laws of arithmetic, at the age of thirteen.

5 × 2 = 10 to which add 3 = 13 which multiplied by 4 gives 52, instead of 5 × 2 = 10 to which add 3 × 4 = 12.[14]

The list of sums and products worked out at the bottom of the page bears out Miss Lamont's belief in Ada's interest and accuracy. Even in these childhood letters we see an intelligent, inquisitive and tenacious pupil, who would go on to study higher mathematics, and enjoy doing so.

Ada's father wanted her to learn music and Italian, but shortly before his death, when Ada was eight, Lady Byron told him how Ada's imagination was 'chiefly exercised in connection with her mechanical ingenuity – her self-invented occupations being the manufacture of ships and boats'.[15]

By the time she was twelve, in 1828, she was writing to her mother about flying:

> Today I have been flying particularly well and I think you will really say I have much improved in that exercise.[16]

The designs for the first flying machines (in Europe at least) are often credited to Leonardo da Vinci, but the first human flights, carried out by the Montgolfier brothers over Paris in 1783, were by balloon, made possible by the discovery of hydrogen some years earlier. Though dreams of ballooning as a means of mass transport faded with the coming of the railways, balloon flights became a popular spectacle – indeed 1828 saw the dramatic death of a Mr Harris while crash-landing his balloon near Croydon, though his passenger, a Miss Stocks, survived. So it is no surprise that the idea of human flight should capture the attention of an imaginative and scientifically inclined twelve-year-old.

Indeed, as well as dreaming of flying, Ada asked Lady Byron to find a book for her on the anatomy of birds and 'had great pleasure in looking at the wing of a dead crow'. She looked

OVERLEAF
15 Letter from Ada Byron, aged twelve, to her mother, describing her ideas about flying.

A MATHEMATICAL CHILDHOOD

Today I have been flying particularly well and I think you ~~would~~ will really say I have much improved in that exercise.

My wings are going on prosperously but do not expect to see a pair of well proportioned wings though they are quite sufficiently so for me to explain to you all my ideas on the subject of flying.

As soon as I have brought flying to perfection, I have got a scheme ~~about a scheme~~ about a, ~~steamengine~~ (pardon my dirty blot), steamengine which, if ever I effect it, will be more wonderful than either steampackets or steamcarriages, it is to make a thing in the form of a horse with a steamengine in the inside so contrived as to move an immense pair of wings, fixed on the outside of the horse, in such a manner

Bifrons near Canterbury, Saturday, 5th February 1828.

My dearest Mammy, I received your pretty little letter this morning & am very glad that you are pleased with my letter and that you gradually improve.

Mrs Paff desires me to express to you her sense of the honour you do her in wishing to be godmother to her eldest daughter. I cannot conceive what you intend me to understand by the word "Whales!" placed quite alone in the middle of your letter with three scratches under the h and a note of admiration at the end, pray explain it's meaning to me. Tomorrow I intend to write to Lady Tamworth — When I sent my letter to you yesterday I had not yet looked at the sermon "On the duties of children to their parents", I read it yesterday evening and like it particularly, there is a great deal in it which relates

16 Sir George Cayley devised various experimental gliders, one of which flew 500 yards in 1849, carrying a ten-year-old boy. The 'governable parachute', described in the 25 September 1852 edition of *Mechanics' Magazine*, was designed to be released from a balloon to glide back to the ground: it was probably not built in his lifetime.

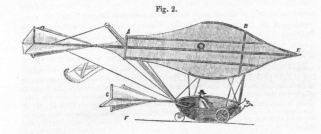

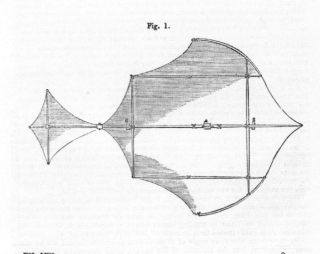

forward to showing to her mother some wings she was making from paper, silk and feathers:

> My wings are going on prosperously but do not expect to see a pair of <u>well</u> proportioned wings though they are quite sufficiently so for me to explain to you all my ideas on the subject of <u>flying</u>.[17]

She hoped to write a book on 'Flyology', and, not content with a glider, she had plans for a flying machine that would be powered by the cutting-edge technology of the age: steam (FIGURE 15). In 1828, a few years after the first commercial railway, she wrote:

> As soon as I have brought <u>flying</u> to perfection, I have got a scheme about a … steamengine which, if ever I effect it, will be more wonderful than either steampackets or steamcarriages, it is to make a thing in the form of a horse with a steamengine in the inside so contrived as to move an immense pair of wings, fixed on the outside of the horse, in such a manner as to carry it up into the air while a person sits on its back.[18]

Ada was aware that this would not be easy, and must be taken step by step:

> This last scheme probably has infinitely more difficulties and obstacles in its way than my scheme for flying but still I should think that it was possible and if I succeed in the flying it will be an encouragement to me to try the horse.[19]

The imaginative Ada was only slightly ahead of her time. The 1840s saw the first manned gliders, designed by the English engineer Sir George Cayley, who had identified the principle that an inclined wing surface will generate lift. The 'aerial steam carriage', a steam-powered flying machine, was patented in 1842, though never built.

3
EARLY INFLUENCES

Lady Byron continued to take care over her daughter's education. Shortly after Ada's thirteenth birthday she sought the advice of her friend Dr William King, the doctor she consulted on her many minor health problems. Lady Byron summed up her daughter's knowledge of mathematics as:

> 'Arithmetic, the first part of Algebra & Paisley's Practical Geometry'. Present information on these subjects confined to some general ideas of the value of numbers, without facility in working them, & to some acquaintance with the problem in the 1st part of Paisley. Great interest in such pursuit.[1]

In the practical spirit of Pestalozzi, the geometry book contained detailed instructions for making technical drawings, of the kind needed by surveyors or engineers, without using any algebra or Euclidean geometry.

In line with her own interests at the time, Lady Byron seemed more interested in her daughter's education in 'Morals, History, and Political Economy', though she also wanted Ada to study 'The principles of Natural History', where she had

17 John Constable, *Landscape with a Double Rainbow*, 1812, oil on paper.

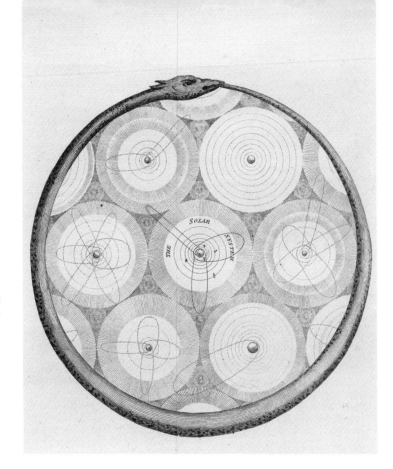

18 Diagram of the solar system from *An Introduction to Astronomy* by John Bonnycastle, 1786. The book, which ran to many editions, presented the principles of astronomy for 'persons who have not acquired a sufficient stock of Mathematical knowledge to read, with satisfaction, the works of Newton, and other eminent writers upon this subject'.

Recent information confined to a few isolated facts. Great interest in the causes of natural phenomena, and their classification. To cultivate that taste would provide occupations calculated to prevent any ill effect from concentration of feeling, or habits of abstraction.[2]

The now elderly William Frend wrote many long letters to Ada's mother about religious matters. He also took an interest in Ada's scientific education, encouraging her to think for herself instead of relying on books:

How does Miss B. come on with her astronomy? The next month toward the end will exhibit Jupiter to her to great advantage & at a reasonable hour. I hope you have a good telescope & it will be an amusing exercise to sketch the planet with his moons & observe the variation of their positions in succeeding nights. She may be fortunate enough to witness a few eclipses & occultations but I would not consult books on the occasion. She may make tolerable guesses at the approaching phenomena & verify them by her own observations.[3]

The early nineteenth century saw increasing popular interest in astronomy, with widely read books and magazines not only describing William Herschel's discoveries, and his giant 40-foot telescope near Windsor, but also celebrating the work of his sister Caroline, the 'lady comet hunter'.

Unfortunately, not long after Frend's letter was written, Ada fell ill with measles, complications from which left her bedridden for many months, interrupting her studies. But she retained her interest in mathematics and science throughout her teenage years. In 1834 she wrote to William Frend (FIGURE 19) to ask him about rainbows:

I am very much interested on the subject just now, but I cannot make out one thing at all, viz: why a rainbow always appears to the spectator to be an arc of a circle. Why is it a curve at all, and why a circle rather than any other curve? I believe I clearly understand how it is that the colours are separated, and the different angles which the different colours must make with the original incident ray. I am not sure that I entirely understand the secondary rainbow.[4]

The mathematical theory of the rainbow was not entirely understood at the time, but Ada speculated correctly:

Fordhook
15.th March
1834

Dear Mr Frend

You have always been very goodnatured to me, & have seemed exceedingly willing to answer any ignorant questions. I shall be very grateful if you will be kind enough the first time you have a few spare moments, to write me a letter about rainbows. I am very much interested on the subject just now, but I cannot make out one thing at all, viz: why a rainbow always appears to the spectator

Is the spectator's eye supposed to be in the centre of the circle of which the arc of the rainbow forms a portion?[5]

She wanted to repair the gaps in her mathematical education, and in March 1834 asked Dr King for further help. 'My wish', she wrote,

> is to make myself well acquainted with Astronomy, Optics &c; but I find that I cannot study these satisfactorily, for want of a thorough acquaintance with the elementary parts of Mathematics.[6]

What she needed, she said, was a course in pure mathematics, by which she meant basic arithmetic, algebra and geometry. Dr King advised a rather old-fashioned course of instruction, and she started work on Euclid with enjoyment. Within a fortnight, she could report that she was

> getting on very well so far, with Euclid. I usually do four new propositions a day, and go over some of the old ones. I expect now to finish the 1st book in less than a week.[7]

A classic result in geometry is Pythagoras' theorem, which states that in a right-angled triangle, the square on the hypotenuse is equal to the sum of the squares on the two adjacent sides. By mid-April, Ada was guessing at her own variation. She wrote to Dr King:

> Can it be proved ... that equilateral triangles being constructed on the sides of a right angled triangle, and also one on the hypotenuse, the sum of the triangles on the sides is equal to the triangle on the hypotenuse? ... It strikes me that it ought to be as demonstrable as when the figures are four-sided & equilateral.[8]

OPPOSITE
19 Letter from Ada Byron to her mother's friend and former tutor, William Frend, 15 March 1834, asking about the mathematics of a rainbow.

OVERLEAF
20 Undated manuscript of mathematical notes by Ada Byron, including a diagram to illustrate her question to Dr King about a variant of Pythagoras' theorem in which triangles replace the usual squares.

EARLY INFLUENCES   27

$$\sin \tfrac{1}{2}a = \sqrt{\tfrac{1}{2}R^2 - \tfrac{1}{2}R\cos a}$$

for $\cos a$ put $\pm\sqrt{R^2 - \sin^2 a}$

$$\sin \tfrac{1}{2}a = \sqrt{\tfrac{1}{2}R^2 \mp \tfrac{1}{2}R\sqrt{R^2 - \sin^2 a}}$$

Let $\sin \tfrac{1}{2}a = \tfrac{1}{2}\sqrt{R^2 + R\sin a} \mp \tfrac{1}{2}\sqrt{R^2 - R\sin a}$

$$\tfrac{1}{2}R^2 \mp \tfrac{1}{2}R\sqrt{R^2 - \sin^2 a} =$$

$$= \tfrac{1}{4}R^2 + \tfrac{1}{4}R\sin a + \tfrac{1}{4}R^2 - \tfrac{1}{4}R\sin a \mp 2 \times \tfrac{1}{2} \times \tfrac{1}{2}\sqrt{R^2 + R\sin a} \times \sqrt{R^2 - R\sin a}$$

$$\mp \tfrac{1}{2}R\sqrt{R^2 - \sin^2 a} = \mp \tfrac{1}{2}\sqrt{R^2 + R\sin a} \times \sqrt{R^2 - R\sin a}$$

$$= \mp \tfrac{1}{2}\sqrt{R^4 - R^2 \sin^2 a} = \mp \tfrac{1}{2}R\sqrt{R^2 - \sin^2 a}$$

$$\mp \tfrac{1}{2}R\sqrt{R^2 - \sin^2 a} = \mp \tfrac{1}{2}R\sqrt{R^2 - \sin^2 a}$$

$(a \pm b)^2 = a^2 + b^2 \pm 2a$

$(a - b) \times \tfrac{}{} $
$= a^2 - b^2$

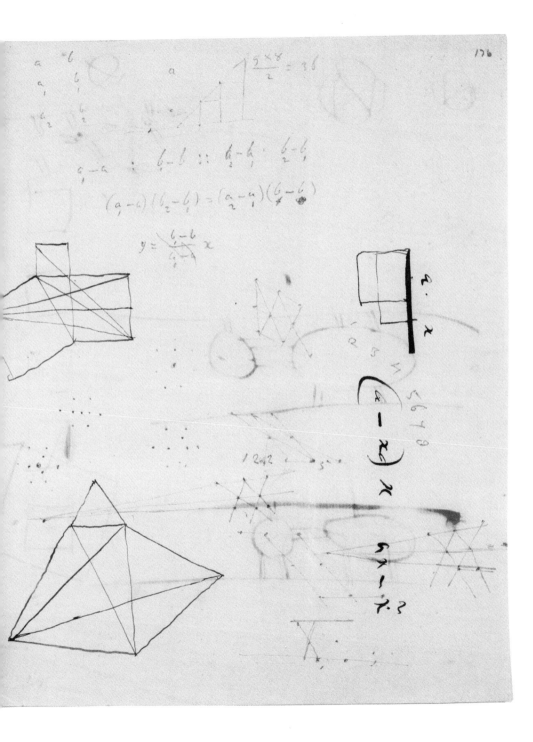

21 Proof of Pythagoras' theorem from Oliver Byrne's 1847 edition of the first six books of Euclid's *Elements*. The book was novel in its use of coloured pictures, rather than text, to demonstrate the proofs.

One way to explain the truth of this is to see that the area of the equilateral triangle is proportional to the area of the corresponding square. So since the area of the square on the hypotenuse is the sum of the areas of the other two squares, the area of the triangle on the hypotenuse is the sum of the areas of the other two triangles.

Dr King had expected Ada to study mathematics in the old-fashioned way he had been taught at Cambridge many years before, by memorizing portions of Euclid to pass exams, and not by trying out new ideas or inventing new proofs. But in just seven weeks she had reached the boundaries of her new tutor's expertise, and he apologized that 'You will soon puzzle me in your studies'.[9]

At about the same time, Ada met Mary Somerville, encouraged by William Frend as a suitable acquaintance 'not more distinguished by her scientific attainments than by her amiable

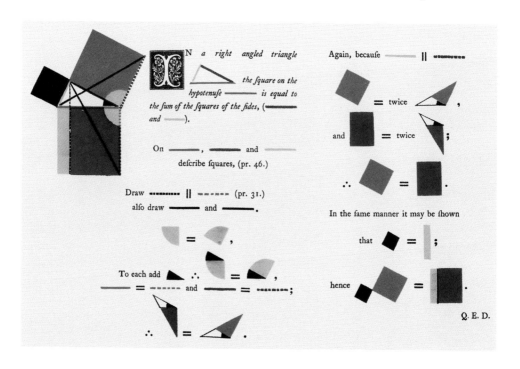

qualities'.[10] Although of the same generation as Dr King, Somerville had a rather more sophisticated mathematical understanding, and was the first practising scientist Ada had met. Somerville and Caroline Herschel were the first female fellows of the Royal Astronomical Society, and perhaps the only women in Britain able to earn money from mathematics. Herschel was paid a royal stipend, and Somerville published widely read textbooks, including her translation of the *Mécanique Céleste* from the French of the mathematician Laplace.

The Somerville family chaperoned Ada Byron in London's scientific and literary society, and in 1833, the year she was presented at court, she met Charles Babbage. Somerville's son, Woronzow Greig, also introduced Ada to his friend Lord William King,[11] a quiet and scholarly man ten years her senior, whom she married in 1835. King shared and encouraged his wife's scientific interests: he was particularly interested in scientific approaches

22 Geometric models by George Adams, c.1750. Ada Byron owned a number of wooden models of this kind, designed to illustrate the principles of geometry.

to agriculture, and, in later life, in elaborate building projects on his country estate. The couple settled down to the life of the landed gentry of the time, occupying a house in London's St James's Square, and eventually two country estates, in Surrey and Somerset. William became the Earl of Lovelace in 1838, and Lord Lieutenant of Surrey in 1840. Three children, Byron, Anabella and Ralph, were born in 1836, 1837 and 1839.

Ada, now Countess of Lovelace, maintained her dedication to mathematics, and continued to correspond with Mary Somerville.

> I now read Mathematics every day & am occupied in Trigonometry & in preliminaries to Cubic & Biquadratic Equations. So you see that matrimony has by no means lessened my taste for those pursuits, nor my determination to carry them on...[12]

She asked Mary Somerville's advice on obtaining wooden models to help visualize facts from spherical trigonometry:

> If three great circles intersect, they will divide the Sphere into eight spherical triangles. Each of the hemispheres into which any one of the circles divides the sphere be divided into four spherical triangles which will be respectively equal side for side and angle for angle with the four triangles of the other hemisphere. Every two of these triangles have an angle in one equal to an angle in the other, and the sides opposite to these angles respectively equal, while the remaining sides and angles are supplemental.[13]

While praising her scientific interests, Mary Somerville also encouraged her in domestic matters, thanking her for an embroidered cap:

> I am very much delighted to have a cap made by you and the more so as it shows [that] we mathematicians can do other things besides studying $x$es and ys.[14]

23 Mary Somerville, self-portrait.

24 Pencil drawing by Frederick Sargent, around 1870, of William, Earl of Lovelace, whom Ada Byron married in 1835.

They had less contact after the Somervilles moved to Italy in 1838, though Woronzow Greig remained a close friend.

Like Lady Byron, the Lovelaces founded a village school, again on the principles of Pestalozzi, and Ada asked her mother who might write an elementary textbook for it. Lady Byron suggested her friend Harriet Martineau, a journalist, and writer on political economy, though she wanted Martineau's book to 'avoid any of her queer theories',[15] and especially disagreed with her support for giving women rights to vote:

1. That they lead women to undervalue the privilege of being exempted from political responsibility. –
2. That by stimulating them to a degree of exertion, intellectually, for which their frames are not calculated, the ability to perform private duties is impaired – Mind is transferred from its legitimate objects.
3. That by mixing imaginary with real grievances, the redress of the latter is retarded.[16]

Ada Lovelace was rich, titled, independent-minded, with a supportive husband, and by now mixing with the elite male and female scientists of the day. But she was still limited by what society and her family expected of her, and still had to rely on the favours of friends for her scientific education.

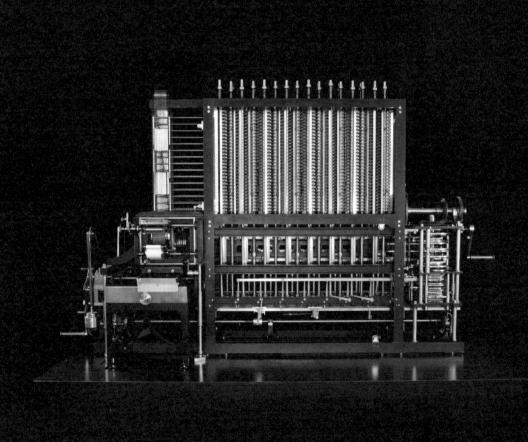

CHARLES BABBAGE AND THE THINKING MACHINE

Charles Babbage grew up in Devon and studied in Cambridge. He was drawn to the newer ideas of European mathematicians, later expressing his dislike of the British scientific tradition in his book *Reflections on the Decline of Science in England, and on Some of Its Causes*. This was just one of the many strongly worded writings, on topics ranging from theology to street musicians, that he produced throughout his life.

Babbage studied a wide variety of industrial processes to understand the organization of mechanized tasks, and his book *The Economy of Machinery and Manufactures* became recognized as a classic account, influencing Karl Marx, among others. The task he particularly wanted to mechanize was the production of printed tables. Nineteenth-century banking, manufacturing and shipping required substantial calculations, for example for insurance or navigation, and relied on a huge variety of these tables, all created by hand. In his autobiography Babbage wrote that he had the idea of producing them 'by steam'[1] while he and John Herschel were checking the entries of an astronomical table.

Babbage's first machine, the Difference Engine (FIGURE 25), was designed to calculate and print successive values of a formula. It used the method of finite differences, which calculates each

25 One of the two working reproductions of Charles Babbage's Difference Engine No. 2, created by the London Science Museum to Babbage's designs between 1985 and 2002.

26 This chart in Charles Babbage's design notation shows the mechanism used in the Difference Engine printer to advance the roll of paper, so as to start a new line of printing.

Aug. 1848.

Motions of Difference Engine N° 2
the Complete Trains and &c

A_2 & A_1	Even Difference Sector		49 & 52	
A_5	A_{11}	" " Figure Wheel	58 & 61	
A_{21}	" " Warning	74 & 77		Circular Motions of the Calculating Axes
A_{22}	Odd Difference Warning	40 & 43		(see Sheet 2 of this Notation)
A_{10}	" " Figure Wheel	24 & 27		
A_6	A_{12}	" " Sector	18 & 18	

A_4	A_9	Even Difference Warning Axis	75 & 82
A_{20}	A_{19}	Odd	44 & 48
A_3	A_8	Even Difference Sector Axis	53 & 57
A_{17}	A_1	Odd	19 & 23
A_{15}	A_{14}	Lock Odd Diff. Figure Wheel	32 & 39
A_{13}	A_{16}	Even	18 & 73
A_{18}	A_{24}	Even Diff. Figure Wheel Axis	62 & 67
A_{23}	A_7	Odd	28 & 31

Vertical Motions of the Calculating Axes
(see Sheet 3 of this Notation)

A_s	Ink Printing Sectors	163 & 167
A_2	A_3	Lock Large Stereotype Sectors	124 & 129
A_2	A_4	" Small "	140 & 144
A_3	A_5	Punch Large Type	192 & 114
A_4	A_7	" Small "	144 & 146
A_6	Lock Vertical Racks	135 — 139
A_8	Printing Sectors	130
A_9	Move Paper for Printing	147 & 157
A_{10}	Roller for Inking	100 & 167
A_3	A_4	Lock Figure Rack	132 & 136

Paper and Printing Apparatus
(See Sheet 4 of this Notation)

Carry on Table axis
" " First Difference Axis
" " Second "
" " Third "
" " Fourth "
" " Fifth "
Carry on Sixth Difference Axis

Carrying Axes
(See Sheet 4 of this Notation)

Advance from Line to Line
Advance from Column to Column
Back from Line to Line
End of Page — Engine Stops itself

Advance from Column to Column
Advance from Line to Line
Back from Column to Column
End of the Page — Engine Stops itself

Motions of the Stereotype Frames.
(See Sheet 5 of this Notation)

27 Daguerreotype of Charles Babbage, around 1850, by the French photographer Antoine Claudet.

value from the previous one by repeated addition. For example, in calculating a list of squares,

$$1, 4, 9, 16, 25 \ldots$$

the value $16 = 4 \times 4$ is calculated from the previous value $9 = 3 \times 3$ by adding $7 = (2 \times 3) + 1$. Written another way,

$$\begin{aligned} 4 \times 4 &= (3 + 1) \times (3 + 1) \\ &= (3 \times 3) + (2 \times 3) + 1 \\ &= 9 + 6 + 1 \\ &= 16 \end{aligned}$$

The general formula for calculating squares is that $(a + 1) \times (a + 1)$ is obtained from $a \times a$ by adding $(2 \times a) + 1$, since

$$(a + 1) \times (a + 1) = (a \times a) + (2 \times a) + 1$$

A similar method works for more complicated formulae.

Using this method meant that Babbage's machine did not have to be able to multiply, only to add and subtract. While this was not the first mechanical calculator, it was unique in being completely automatic. Once it had been set up to calculate the values of the required formula, the only human intervention needed was to turn the large handle at the side, and to change over the wax tablets on which the results were printed. These tablets could then be used to make plates for printing the final tables. Working with the talented engineer Joseph Clement, Babbage experimented with many ingenious designs for the mechanical components of his machine. The numbers were stored using metal wheels designed always to move to one of ten fixed positions, with a clever mechanism to recover from mechanical errors. The final design held 8 numbers of 31 digits, in 8 stacks around 3 feet high, each containing 31 such wheels. If a calculation required a large number of parts to move simultaneously, the machine would be subject to huge forces, so it anticipated such cases and did some of the calculations in advance. The components, and there were 8,000 in all, needed to be accurately made, and reproducible in quantity. Babbage even invented his own 'mechanical notation', which described complex machinery by combining a collection of symbols representing basic components (FIGURE 26).

Babbage raised substantial government funds of over £17,000, at a time when a railway engine cost around £800, to build his Difference Engine, but lengthy wrangling with Clement over contracts meant that by 1832 all he had to show after ten years of work was a large collection of components and a small demonstration

28 Lady Byron describes a tour of factories in 1834, including the child employees, the role of trade unions and the punch cards controlling the machinery.

original system has not been disturbed by
cultivators of the soil is the proprietor. — This
does exist so far, as to make a certain po[rtion]
the Landlord's by right. State the Land[lord]
cannot sell the Estate. The benefit[s]

direction, so that the side bd will now
be uppermost, instead of the side ac,
and the ball retraces its course till
it arrives at a, and the minute
is then completed. There is a clock
face above, on which the hand
marks these twenty revolutions — but by a
sudden motion from one side of the
dial to the opposite one, instead
of a gradual progress. — The Clock will
not go accurately — it errs by 5 minutes
a day — However, as Mr Stoutt observes,
it may be perfected in time. —
Lawrence's unfinished portrait of Laura
was amongst the modern collection — Martin's
picture of the Waters of Oblivion which like the
Eruption of Vesuvius by the same Painter, which
I saw at Stowe, gives me more pleasure than
its merit as a work of art can justify. — One is
placed completely beyond the region of possibilities in
looking at the former, and in the region of terror
& sublimity
in looking at the latter — There were two busts of
Mr J. Stoutt & his deceased Brother — a philanthropist —
The cause of benevolence in the head of the latter,
was unmistakable — but why has Mr Stoutt such an ordinary pate!

July 17th — Saw the ribbon Ma[nufactory]
Coventry. The Patterns, wi[th exception]
of the very simple ones, a[re produced]
by stamped pieces of pape[r]

[punch card illustration]

Some of the holes are fil[led others]
left open — and this the [machine]
pass, which afterwards [guides the]
threads. — The highest wag[es paid to]
the children employed in [tying]
ends of the silk which [forms the]
surface of the ribbons [— work]
from 6 in the morning [till 7 in the]
evening. ½ hour allow[ed for breakfast]
1 hour for dinner — ½ f[or tea]
in the morning, set in [going]
to Sunday school generally [taken]
as early as 7 years old [— not]
considered alarming — but th[e]
frequent strikes. — It was [said]
The rules of a sick fund were [read]
of the Walls. —
At Ashby de la Zouch, saw
Spar Manufactory — where to[ols]
are used — went out to det[ach]
July 18. Saw China Manufa[ctory]
Infirmary — where I was intere[sted]

Fordhook
June 21.

Dear Dr King — I begin to-day a series of not very distant excursions, which would prevent my receiving an impromptu visit from you. I hope I may be in the way when Mrs King returns — will you tell her so? — The week after next I shall be in the neighbourhood of Cambridge — To-day I am going to Beulah Spa — and Ada is going to the Queen's Ball! — I am better pleased with my destination than she is with hers, tho' she has some pleasure in expecting to meet Babbage at the palace! — We both went to see the thinking machine (for such it seems) last Monday. It raised several Nos. to the $2^{nd} \& 3^{rd}$ powers, and extracted the root of a Quadratic Equation. — I had but faint glimpses of the principles by which it worked — Babbage said it had given him notions with respect to general laws which were never before presented to his mind. — For

29 Letter from Lady Byron to her friend Dr King, 21 June 1833, describing Charles Babbage's 'thinking machine'.

model. This is now in the London Science Museum, along with a complete Difference Engine that was finally constructed to one of Babbage's later designs in 1992, and that works perfectly.

Babbage never wrote an account of his machine, but it was publicized through the lectures and writings of his friend Dionysius Lardner, and through Babbage's own Saturday evening parties. Charles Dickens, Isambard Kingdom Brunel and Charles Darwin were regular attendees, and saw Babbage's demonstrations of the Difference Engine. Also on display was his collection of mechanical automata, including a famous Silver Lady which

> used an eye-glass occasionally, and bowed frequently, as if recognizing her acquaintances.[2]

30 The printing mechanism from the second reproduction of Charles Babbage's Difference Engine.

Lady Byron and the seventeen-year-old Ada attended in June 1833, and Lady Byron afterwards wrote to Dr King that

> We both went to see the thinking machine (for such it seems) last Monday. It raised several Nos to the 2nd & 3rd powers, and extracted the root of a Quadratic Equation. I had but faint glimpses of the principles by which it worked.³

Ada was captivated, went to Lardner's lectures to learn more, and studied Babbage's engineering drawings. Many years later her mother's friend Sophia De Morgan, the daughter of William Frend, recalled:

> While other visitors gazed at the working of this beautiful instrument with the sort of expression, and I dare say the sort of feeling, that some savages are said to have shown on first seeing a looking-glass or hearing a gun if, indeed, they had as strong an idea of its marvelousness, Miss Byron, young as she was, understood its working, and saw the great beauty of the invention.⁴

The young Ada's interest in mechanical things had been developed by her mother through tours of factories: following their visit to a ribbon factory in 1834, the reform-minded Lady Byron described the young children who worked there, the role of trade unions in improving working conditions, and the punch-card mechanism used to control the machinery (FIGURE 28).

All her life Ada remained fascinated by the latest inventions. She wrote enthusiastically about Charles Wheatstone's telegraph:

> sometimes friends hold conversations from one terminus to the other; that one can send for anyone to speak to one. ... I might desire a tradesman to go directly to the Nine Elms Station and might discuss an order with him about goods etc. Wonderful agents and invention!⁵

Her intense study of Babbage's engine reinforced her desire to learn more advanced mathematics, and after many requests to friends, in 1840, at the age of twenty-four, she finally found a tutor.

OVERLEAF
31 Letter from Ada Lovelace to Augustus De Morgan, 19 February 1841, with questions about the 'Calculus of Finite Differences', the mathematical principle underlying the Difference Engine.

the top of page 79:

It is very clear that the law for the Co-ef being proved for u_n, and for Δu_n, fo immediately & easily for u_{n+1}, or $u_n + \Delta $

But if we now wish to establish for u_{n+2}, we must prove it true n only for u_{n+1}, but also for Δu_{n+1}

To retrace from the beginning: the object in the first half of page 79 evide is to prove firstly, that any order of $ $ say u_n can be expressed in terms of u a Series of all the Differences of u, Δu $\Delta^2 u$, $\Delta^3 u$, -------------- $\Delta^n u$; Secondly, that the Co-efficients for this Ser follow the law of those in the Binomial Th

Now the first part is evident from th law of formation of the Table of Difference Since all the Differences Δu, $\Delta^2 u$, Δ^3 are made out of u, u_1, u_2 &c., it obvious that by exactly retracing & reaso the proofs, we can make u, u_1, u_2 &

...of Δu, $\Delta^2 u$, $\Delta^3 u$ &c.

...on the second part of the above; if we
...show that the law for the Co-efficients
...do good up to a certain point, say u_4;
...also, that being true for any one
...ue, it must be true also for the next
...ue too; the demonstration is effected for
... values.

... the fact is shown that it is true up
... u_4. (I must not here enquire why the
... is so. That is I suppose not your
... ranging, or any part of your affair).—
... is shown that the two parts u_3, Δu_3 of
... u_4 is made up are under this law,
... therefore that u_4 is so. And next it is
... ... that any other two parts u_n, Δu_n
... ing under this law, their sum u_{n+1}
... ust be so. But this proves nothing
... a continued succession. u_{n+1} being
... der this law does not prove that Δu_{n+1}
... under it, & therefore that u_{n+2} is under it.

5
LEARNING MATHEMATICS WITH PROFESSOR DE MORGAN

Ada's new tutor was Augustus De Morgan, the Professor of Mathematics at University College London, which had opened in 1828. Although one of the leading British mathematicians of the time, he is viewed now as a minor figure, and the influence of his European contemporaries has been more long-lasting. He is usually remembered for 'De Morgan's Laws', a fundamental principle of the logic of software and circuit design, but he was also a prolific and witty writer on all areas of mathematics, publishing numerous research papers, popular articles, book reviews, encyclopedia entries and textbooks.

De Morgan, like Babbage, disliked the old-fashioned approach to mathematics of his studies at Cambridge, and in his own teaching wanted his students to understand first principles rather than working through standard exercises. He was highly regarded as a teacher, and alongside his university classes also took private pupils – all male, so far as we know, apart from Ada Lovelace. He initially taught her as a favour to his wife Sophia. Although his male pupils would have paid for tuition, it seems likely that Ada Lovelace did not. His letters often include thanks for presents of game from the Lovelace country estates (FIGURE 33), while hers frequently apologize for taking up his time.

32 Augustus De Morgan teaching, drawn by a student at University College London on 17 March 1865.

33 Letter from Augustus De Morgan to Ada Lovelace, September 1840, explaining that many different curves can be drawn passing through a set of points, and thanking her for 'very good partridges' received from her country estate.

34 Daguerreotype of Ada Lovelace, c.1842, by the French photographer Antoine Claudet.

By the summer of 1840 Ada Lovelace was reporting to her mother that 'I have written to the professor twice'[1] and that she was pleased with her progress. The arrangement was simple: she took what we would now call a correspondence course with De Morgan, consisting of guided independent reading and a variety of exercises. When Ada was in London, the professor and his pupil would sometimes meet face-to-face, either at the Lovelace's mansion in St James's Square or at the De Morgan family home about a mile away, in Gower Street in Bloomsbury. Although often combined with social visits, these meetings would usually be prompted by some difficulty Ada was having which she felt could be best explained in person. As she put it: 'I should prefer <u>seeing</u> you to writing, on this occasion.'[2] In one of his replies De Morgan extended an invitation to both Ada 'and Lord Lovelace too if he be not afraid of the algebra'.[3] But more often than not, they simply conversed via letter, with his pupil sending occasional packets of worked mathematical problems for De Morgan to look over, accompanied by further apologies for taking up his time: 'I am afraid you

...quired the
...ng belong to
... very difficult

... you a glimpse
...tion to a
...t paper
... points P Q R

let $x = a, \; y = A$
$ x = b, \; y = B$
$ x = c, \; y = C$

$B \dfrac{(x-c)(x-a)}{(b-c)(b-a)} + C \dfrac{(x-a)(x-b)}{(c-a)(c-b)} + \left\{ \begin{array}{l} \text{any function of } x \text{ which} \\ \text{does not become infinite} \\ \text{when } x = a, \text{ or } b, \text{ or } c \end{array} \right\} \times (x-a)(x-b)(x-c)$

...umber of equations which you will find to satisfy the condition

...re very good
...rom Ockham
...nces to Lord Lovelace
...rs very truly
DeMorgan

I have heard of Lady Byron by
Mr Phillott who left her safe
at Fontainebleau

35 Ada Lovelace's letter to Augustus De Morgan, January 1841, in which she compares mathematical expressions to 'sprites & fairies'.

impossible *identicity* of form exceedingly *dissimilar* at first sight, is I think one of the chief difficulties in the early part of mathematical studies. I am often reminded of certain sprites & fairies one reads of, who are at one's elbow in one shape now, & the most minute a form the most dissimilar and uncommonly deceptive troublesome & tantalizing as the mathematical sprites fairies sometimes; like the types I have found for t[hem] in the world of Fiction

I will now go to the question
I delayed asking before:
In the development of the
Exponential Series
$$a^x = 1 + (\log a)x + \frac{(\log a)^2 x^2}{2} + \&c,$$
and the Logarithmic Series
$$\log a = (a-1) - \frac{1}{2}(a-1)^2 + \&c$$
deduced from it; I object
to the necessity involved of
supposing x to be <u>diminished
without limit</u>, — a supposition
obviously [?] necessary to the completion
of the Demonstration. It has
struck me that though this
supposition leaves the Demonstration
& Conclusions perfect for the
cases in which x <u>is</u> supposed
to <u>diminish</u> without limit, yet

will indeed say that the office of my Mathematical Counsellor or Prime-Minister is no joke.'[4] This arrangement seems to have worked well. Ada wrote in July of 1840: 'I think the professor suits me exceedingly.'[5]

Only part of the eighteen-month correspondence between them survives: forty-two lengthy letters from Ada Lovelace and twenty replies, often much shorter, from De Morgan. The letters form a fascinating record of the interaction between a gifted and sympathetic teacher and an intelligent and motivated student. Lovelace was studying at roughly the level of De Morgan's second-year university classes, and the letters show his judgement in balancing this more interesting and advanced material with the basic study of geometry and algebra needed to repair the gaps in his pupil's knowledge.

All mathematics students, even the most talented, find some things difficult when they first encounter them. The correspondence is unusual in providing a record of such difficulties, and the way the student learns from overcoming them. In one letter, Lovelace realized that a problem can be solved by making a substitution.[6]

> It had not struck me that, calling $(x + \vartheta) = v$, the form $\dfrac{(x + \vartheta)^n - x^n}{\vartheta}$ becomes $\dfrac{v^n - x^n}{v - x}$

She reflected that

> the curious <u>transformations</u> many formulae can undergo, the unexpected & to a beginner apparently <u>impossible identity</u> of forms exceedingly <u>dissimilar</u> at first sight

might be the cause of

> the chief difficulties in the early part of mathematical studies.

She also noted, somewhat whimsically:

> I am often reminded of certain sprites & fairies one reads of, who are at one's elbow in one shape now, & the next minute in a form most dissimilar; and uncommonly deceptive, troublesome & tantalizing are the mathematical sprites & fairies sometimes; like the types I have found for them in the world of Fiction.

Wrestling with another problem, she wrote:

> I do not know when I have been so tantalized by anything, & should be ashamed to say how much time I have spent upon it, in vain.[7]

Describing certain equations as 'complete Will-o'-the-Wisps to me', she lamented:

> The moment I fancy I have really at last got hold of something tangible & substantial, it all recedes further & further & vanishes again into thin air.[8]

In Figure 36 we see Ada working out for herself how to draw the 'equation to a curve', which we would now call the graph of the function $y = x^2$. Her earlier lessons had included plenty of geometry, but nothing that connected algebra and geometry like this. Taking successive values of x, namely ¼, ½, ¾, 1, 3/2, 2, ..., she has multiplied them by themselves to get 1/16, ¼, 9/16, 1, 9/4, 4, ..., and drawn vertical lines of these sizes on her diagram.

She then observed that 'By writing the successive perpendicular extremities 1/16, ¼, 9/16, 1, 9/4, 4, &c &c, a curve appears to be produced'[9] which is the visual representation of the equation $y = x^2$. But, she objected,

> This does not appear to me to hold good, as I should have said that the perpendicular straight lines are the

LEARNING MATHEMATICS

as an exercise in algebraical work.

With regard to the logarithms, in the first place, Bourdon is too long. If you will look at the chapter in my algebra, you will find it shorter.

In the equation
$$a^b = c$$
b is called the logarithm of c to the base a. This is the meaning of the term. But for convenience the series $1 + 1 + \frac{1}{2} + \frac{1}{2 \times 3} + \frac{1}{2 \times 3 \times 4} + \&c\ ad\ inf.$ or $2.7182818\ldots$ (called ε) is the base always used in theory; while when assistance in calculation is the object, 10 is always the base; thus if
$$\varepsilon^x = y \qquad x \text{ is the logarithm of } y$$

Thus a is by definition synonymous with = Unit of length

RIGHT
36 Ada Lovelace's question, August 1840, about her visual representation of the equation $y = x^2$, and De Morgan's reply.

LEFT
37 Augustus De Morgan's explanation of logarithms, August 1840.

ε being 2.7182818...

e successive values of x, $\frac{1}{4}$, $\frac{1}{2}$, $\frac{3}{4}$, 1, $\frac{3}{2}$, 2, $\frac{5}{2}$, 3, 4
selected & the corresponding functions x^2, represented
perpendicular lines drawn from the extremities of the
e lines representing x.

uniting the successive perpendicular extremities $\frac{1}{16}$, $\frac{1}{4}$,
1, $\frac{9}{4}$, 4 &c &c, a curve appears to be produced
lines 13 & 14, this curve (according to my interpretation)
ded to as "the representation of a function, or functions
does not appear to me to hold good, as I
d have said that the perpendicular straight lines
the representations of the functions, & I do not see any
e relation that the uniting curve holds to them.

The precise relation is that this one curve, and no
other, belongs to $y = x^2$. Of course there could be no
visible relation unless to a person whose eye
was so good a judge of
length that he could see the
ordinate increasing with
the square of the abscissa.

representation of the function, & I do not see any <u>precise relation</u> that the existing curve holds to them.[10]

A final passage contains De Morgan's response:

> The precise relation is that this one curve, and no other, belongs to $y = x^2$. Of course there could be no <u>visible</u> relation unless to a person whose eye was so good a judge of length that he could <u>see</u> the ordinate [i.e. the vertical line length, or y] increasing with the square of the abscissa [i.e. x^2].[11]

In another letter to Ada (FIGURE 37), De Morgan explains logarithms. In doing so he also introduces the related number ε, which we would write as e today. This is the number given by the formula

$$1 + \frac{1}{1} + \frac{1}{1 \times 2} + \frac{1}{1 \times 2 \times 3} + \frac{1}{1 \times 2 \times 3 \times 4} + \cdots$$

which equals 2.7182818.... It is used when studying quantities which grow very quickly or decline very slowly, for example population growth and radioactive decay, and thus for the mathematics behind epidemiology and carbon dating. Logarithms are used in the measurement of, among other things, economic growth rates, the strength of earthquakes and decibel levels.

In their intense exchange of letters we see Ada not only learning new material, but also learning *how* to learn: going more slowly, learning from mistakes and having a realistic expectation of what can be done. She was an enthusiastic pupil, and when the lessons started wanted to make rapid progress. As she wrote in September 1840:

> I could wish I went on quicker. That is, I wish a human head, or <u>my</u> head at all events, could take in a great deal <u>more</u> & a great deal more <u>rapidly</u> than is the case; and if

I had made my own head, I would have proportioned its wishes & ambition a little more to its capacity. … When I compare the very little I do, with the very much – the infinite I may say – that there is to be done; I can only hope that hereafter in some future state, we shall be cleverer than we are now.[12]

De Morgan provides steady but reassuring words of caution:

> never estimate progress by the number of pages. You can hardly be a judge of the progress you make, and I should say that it is more likely you progress rapidly upon a point that makes you think for an hour, than upon an hour's quick reading, even when you feel satisfied. That which you say about the comparison of what you do with what you see can be done was equally said by Newton when he compared himself to a boy who had picked up a few pebbles from the shore … so that you have respectable authority for supposing that you will never get rid of that feeling; and it is no use trying to catch the horizon.[13]

By November, his pupil was taking this advice. She reported to her mother: 'I work on very slowly. This Mr De Morgan does not wish otherwise. On the contrary he cautioned me against a wish I had at one time to proceed rather too rapidly.'[14] By Christmas, she had changed her mind about how to learn:

> I used once to regret these sort of errors & to speak of time lost over them. But I have materially altered my mind on this subject. I often gain more from the discovery of a mistake of this sort, than from 10 acquisitions made at once & without any kind of difficulty.[15]

And the new year of 1841 found her writing excitedly to Lady Byron: 'I go on most delightfully with Mr De Morgan. … No two people ever suited better.'[16]

See De Morgan's Differential Calculus — page 102

To integrate $\int fx \times \frac{dx}{dt} dt$, x being a function of t, and the integration to be with respect to t, from $t = b$ to $t = b + k$

Let $x = \varphi t$, and $\frac{dx}{dt} = \varphi' t$

Whence $\int fx \times \frac{dx}{dt} dt = \int f(\varphi t) \times \varphi' t \times dt$

And $\int_t^{b+k} fx \times \frac{dx}{dt} dt = \int_t^{b+k} f(\varphi t) \times \varphi' t \times dt$

Let $f_1 x$ = primitive function of fx

Firstly: for $\int fx \frac{dx}{dt} \cdot dt$, we may substitute $\int \frac{df_1(\varphi t)}{dt} dt$

or for $\int_t^{b+k} fx \frac{dx}{dt} \cdot dt \cdots \cdots \cdots \int_t^{b+k} \frac{df_1(\varphi t)}{dt} dt$

Since $fx \frac{dx}{dt} = \frac{df_1 x}{dx} \cdot \frac{dx}{dt} = \frac{df_1(\varphi t)}{d(\varphi t)} \cdot \frac{d(\varphi t)}{dt}$ which last is

by the Rules of Differentiation $= \frac{df_1(\varphi t)}{dt}$ or $\frac{df_1 x}{dt}$

Secondly: by pages 100 & 101, (1) $\varphi x + C = \int_a^x \varphi' x \cdot dx = \int_a^x \frac{d\varphi x}{dx} dx$

(2) And $\varphi(a+h) - \varphi a = \int_a^{a+h} \varphi' x \cdot dx = \int_a^{a+h} \frac{d\varphi x}{dx} dx$

~~~~~~

Therefore if in $\int_t^t \frac{df_1(\varphi t)}{dt} dt$, we consider $f_1(\varphi t)$ as equivalent to φx in (1); t as equivalent to x; b as equivalent to a; we have $\int_b^t \frac{df_1(\varphi t)}{dt} dt = f_1(\varphi t) + C$

And Similarly $\int_b^{b+k} \frac{df_1(\varphi t)}{dt} dt = f_1 \varphi(b+k) - f_1 \varphi b$, derived from (2)

Whence the rest, as in page 103

what does equivalent mean? it is supposed that a and $a+h$ are the values of x when $t = b$ or $b+k$. a is not the equivalent of b
but if $x = \varphi(t)$
$a = \varphi b$
$a + h = \varphi(b+k)$

BECOMING A MATHEMATICIAN

De Morgan's monumental 800-page *Differential and Integral Calculus*, originally published in twenty-five parts between 1836 and 1842, formed the main part of Ada Lovelace's studies. The book set out to be a comprehensive approach to all the mathematics needed to study problems in science. It emphasized the development of mathematics from first principles, with new concepts, and proofs of theorems, presented in great detail. At its core is the first account in English of the rigorous approach to the calculus developed by the French mathematician Augustin-Louis Cauchy. The fundamental idea is that of a limit. For example, the sequence of numbers

    0.9; 0.99; 0.999; 0.9999

and so on, gets closer and closer to 1, whereas the sequence of numbers

    9; 99; 999; 9999

just gets larger and larger. Making this idea of a limit precise, and understanding which sequences of numbers do and do not have a limit, is at the heart of making the rest of calculus precise as well, and of understanding concepts such as differentiation and

38 This document shows Ada beginning to get to grips with the subject of integration, an important topic in calculus. De Morgan's comments at the bottom of the manuscript ask her to reflect on the meaning of the word 'equivalent'.

integration. The great length of De Morgan's book results from its comprehensive approach, and the author's style, which, like his letters to his pupil, is full of asides, explanations, advice to the student and historical references.

By the summer of 1841 Lovelace was making substantial progress. She was more able to identify the cause of her difficulties, writing to De Morgan of

> my Algebra wits ... not having been quite proportionally stretched with some of my other wits.[1]

De Morgan was suggesting more demanding homework, not just routine exercises to give practice in algebra, but also working through sections of his book to make sure she understood the underlying principles. His comments show him encouraging his pupil to think more deeply, for example not just correcting an error but giving hints so she can work out her mistake for herself (FIGURE 38).

Lovelace devoted herself to her studies with ferocious intensity, but she also found time for singing, playing the harp, guitar and piano, and going to the opera. She sometimes complained of the demands of her three small children, sending them, with their nursemaid, to one of her other houses, or to the care of Lady Byron. She went skating; she enjoyed riding and horses, and later in life followed horse-racing, apparently running up substantial debts. With lengthy interruptions due to illness, travel and family matters, she appears to have spent around twelve months studying with De Morgan, and in that time developed her knowledge and understanding of mathematics and mathematical thinking.

In his autobiography, Charles Babbage recalled that, during the composition of Lovelace's paper on his Analytical Engine, she had 'detected a grave mistake which I had made'.[2] During the correspondence with De Morgan we see her obsessive attention to detail, and her desire to work things out from first principles.

ON THE BINOMIAL THEOREM.

$$(1+x)^n \times (1+x)^m = (1+x)^{n+m}$$

or
$$\varphi n \times \varphi m = \varphi(n+m)$$

but when n is a whole number $(1+x)^n$ is the series in question; therefore, calling the above series φn, we have, *when n is a whole number*,

$$\varphi n \times \varphi m = \varphi(n+m)$$

or
$$\left(1 + nx + n.\frac{n-1}{2}x^2 + n.\frac{n-1}{2}.\frac{n-2}{3}x^3 + \&c.\right)$$
$$\times \left(1 + mx + m\frac{m-1}{2}x^2 + m\frac{m-1}{2}\frac{m-2}{3}x^3 + \&c.\right)$$
$$= 1 + (m+n)x + (m+n)\frac{m+n-1}{2}x^2 + (m+n)\frac{m+n-1}{2}\frac{m+n-2}{3}x^3 + \&c.$$

This may be verified to any extent we please, by actual multiplication; for the two first series multiplied together give

$$1 + (m+n)x + \left(n.\frac{n-1}{2} + nm + m\frac{m-1}{2}\right)x^2$$
$$+ \left(n.\frac{n-1}{2}\frac{n-2}{3} + n.\frac{n-1}{2}m + nm.\frac{m-1}{2} + m.\frac{m-1}{2}\frac{m-2}{3}\right)x^3 + \&c.$$

But $n.\dfrac{n-1}{2} + nm + m.\dfrac{m-1}{2} = \dfrac{n^2 - n + 2nm + m^2 - m}{2}$

$$= \frac{(n+m)^2 - (n+m)}{2} = (n+m)\frac{n+m-1}{2}$$

$$n.\frac{n-1}{2}\frac{n-2}{3} + n.\frac{n-1}{2}m + nm\frac{m-1}{2} + m\frac{m-1}{2}\frac{m-2}{3}$$
$$= \frac{n^3 - 3n^2 + 2n + 3n^2m - 3nm + 3nm^2 - 3nm + m^3 - 3m^2 + 2m}{2 \times 3}$$
$$= \frac{(n+m)^3 - 3(n+m)^2 + 2(n+m)}{2 \times 3} = (n+m)\frac{n+m-1}{2}.\frac{n+m-2}{3}$$

and so on. We now lay down the following principle: *When an algebraical multiplication, or other operation, such as has hitherto been defined, can be proved to produce a certain result in cases where the letters stand for whole numbers, then the same result must be true when the letters stand for fractions, or incommensurable numbers, and also when they are negative.* For we have never yet had occasion to distinguish results into those which are true for whole numbers, and those which are not true for whole numbers; but all processes have

40 Portrait of Sir William Rowan Hamilton by Sarah Purser, c.1894.

You know I always have so many metaphysical enquiries & speculations which intrude themselves, that I never am really satisfied that I understand <u>anything</u>; because, understand it as well as I may, my comprehension <u>can</u> only be an infinitesimal fraction of all I want to understand about the many connexions & relations which occur to me, <u>how</u> the matter in question was first thought of or arrived at, &c., &c.³

As she became more confident, she was able to point out not just misprints, but also deeper problems in De Morgan's textbooks. Perhaps the most striking example occurred while reading his *Elements of Algebra* (FIGURE 39), when she wrote:

> I am not at all sure that I like the assumption in the last paragraph of page 212. It seems to me somewhat a large one, & much more wanting of proof than many things which in Mathematics are rigorously & scrupulously demonstrated.[4]

De Morgan responded by defending the assumption, the 'Principle of the Permanence of Equivalent Forms', widely held to be true at the time. Roughly speaking, this stated that if a theorem about numbers is true for whole numbers, then it is true for all numbers.

For Lovelace and De Morgan, 'numbers' included what we now call *real numbers* – that is, whole numbers, fractions, and quantities like π, ε, or the square root of 2. *Real numbers* obey standard rules; so, for example, if you multiply two of them together you get the same answer regardless of the order. So

$2 \times 3 = 3 \times 2 = 6.$

But 'numbers' also included a newer kind of number, developed in the eighteenth century, the so-called *complex numbers*. These are built up from the *imaginary number* called *i*, a new number whose square is –1, so that

$i^2 = -1.$

A *real number* can be drawn as a point on a one-dimensional number line, but a *complex number*, like 3 + 4*i*, has to be drawn in two dimensions, thus:

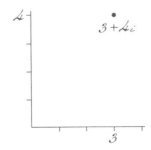

In 1841, after reading an article he had written on complex numbers, Ada Lovelace wrote to De Morgan that

> It cannot help striking me that this extension of Algebra ought to lead to a further extension similar in nature, to Geometry of Three Dimensions; & that again perhaps to a further extension into some unknown region, & so on ad-infinitum possibly.[5]

This was a strikingly accurate prediction, and indeed two years later the Irish mathematician Sir William Rowan Hamilton announced the discovery of *quaternions*, a three-dimensional version of complex numbers. Later in the century others resolved the question for higher dimensions. Quaternions were unusual numbers because they did not obey all of the normal rules of arithmetic; for example, if you multiply certain quaternions together in different orders you can get different (but still correct) answers.

The discovery of quaternions also showed that the 'Principle of Permanence', which Ada had criticized some time before, was wrong. This is a striking example of her ability to spot mathematical difficulties which had escaped the experts.

Quaternions in turn gave rise to vectors, which are now used to study questions from physics in three-dimensional space, and are indispensable in engineering – supersonic flight, satellite navigation and computer chip design. Although Ada Lovelace cannot take credit for their discovery, or for these recent developments, her initial prediction showed remarkable insight.

De Morgan's respect for his pupil was growing, and they were soon discussing some of his more controversial ideas. Mathematicians call the formula

$$1 + x + x^2 + x^3 + \ldots$$

an *infinite series*, because in principle the terms

$$x, \ x^2, \ x^3, \ \ldots$$

go on for ever.

41 Letter from Ada Lovelace to Augustus De Morgan, 19 September 1841, on complex numbers, and her speculation on extending them to a 'Geometry of Three-Dimensions; & that again perhaps to a further extension into some unknown region, & so on ad-infinitum possibly'.

Right-Angle and the angle made by sin.α with the Unit Line. I enclose you an explanation I have written out (according to the Definitions of the Geometrical Algebra), of the two formulae for the Sine and Cosine. I am at work now on the Trigonometrical Chapter of the Differential Calculus. (of the Calculus)

I do not agree to what is said in page 119, that results would be the same whether we worked algebraically with forms expressive of quantities or not. It is true that the form $a + \sqrt{m} - \sqrt{n}$, if (-1) be substituted for m and n, the results come out the same as if we work with a only. But were the form $a \times \sqrt{m}$, $a - \sqrt{m}$, $a \times \sqrt{m}$, or fifty others one can think of, surely the substitution of (-1) for m will not bring out results the same as if we worked with a only; and in fact can only do so when the impossible expression is so introduced as to neutralize itself, if I may so speak. I think I have explained myself clearly.—

It cannot help striking me that this extension of Algebra ought to lead to a further extension similar in nature, to Geometry of Three Dimensions; & that again perhaps to a further extension into some unknown region, & so on ad-infinitum possibly. And that it is especially the consideration of an angle = $\sqrt{-1}$, which should lead to this; a symbol which when it appears, seems to me in no way more

$x < 1$. But if $x > 1$, say $x = 2$, we have
$$\frac{1}{1-2} \text{ or } -1 = 1 + 2 + 4 + 8 + 16 + \&c$$
which, arithmetically considered is absurd. But nevertheless -1 and $1 + 2 + 4 + 8 + \&c$ have the same properties

This point is treated in the chapter on the meaning of the sign $=$.

My wife desires to be kindly remembered

I remain yours very truly

A De Morgan

69 Gower St.
Thursday Ev.^{g} Octr 15/40

It is fair to tell you that the use of divergent series is condemned altogether by some modern names of very great note. For myself I am fully satisfied that they have an <u>algebraical truth</u> wholly independent of arithmetical considerations; but I am also satisfied that this is the most difficult question in mathematics

It can be proved by the methods of calculus that if we substitute a value for x which is less than 1, then the series is equal to

$$\frac{1}{1-x}.$$

If

$$x = \frac{1}{2}$$

for example, then the series is equal to

$$1 + \frac{1}{2} + \frac{1}{4} + \frac{1}{8} + \ldots$$

and has the value

$$\frac{1}{1-\frac{1}{2}} = 2.$$

This kind of infinite series is known as *convergent*. De Morgan, however, was more interested in what would happen if you played with the formula and substituted values of x which are greater than 1. For example, if you substitute $x = -1$ you get the strange formula

$$1 - 1 + 1 - 1 + \ldots = \tfrac{1}{2}$$

and if you substitute $x = 2$ you get another strange formula

$$1 + 2 + 4 + 8 + \ldots = -1.$$

These strange formulae are known as *divergent series*. At this time, mathematicians thought divergent series were meaningless and uninteresting. They are now known to be of great value in mathematics. String theory, which explains the physics of the first moments of the universe, makes use of the peculiar formula

$$1 + 2 + 3 + 4 + 5 + \ldots = -\tfrac{1}{12}.$$

42 Letter from Augustus De Morgan to Ada Lovelace, 15 October 1840, introducing her to his controversial ideas on divergent series, which he describes as 'the most difficult question in mathematics'.

De Morgan believed, correctly, that, while they needed to be handled with care, divergent series could sometimes be useful. As he wrote to his pupil:

> It is fair to tell you that the use of divergent series is condemned altogether by some modern names of very great note. For myself I am fully satisfied that they have an <u>algebraical</u> truth wholly independent of arithmetical considerations: but I am also satisfied that this is the most difficult question in mathematics.[6]

In the 1842 preface to his calculus textbook De Morgan justified introducing his students to the 'uncertain boundaries of known mathematics', by claiming 'the way to enlarge the settled country has not been by keeping within it, but by making voyages of discovery'.[7] Perhaps he was thinking of Ada Lovelace when he wrote:

> the few in this country who pay attention to any difficulty of mathematics for its own sake come to their pursuit through the casualties of taste or circumstances and the number of such casualties should be increased by allowing all students whose capacity will let them read on the higher branches of applied mathematics.[8]

Ada's letter to De Morgan dated 21 November 1841 shows her growing mathematical knowledge and understanding, mentioning his articles on 'Series', 'Operation', 'Differences of Nothing' and the 'Numbers of Bernoulli', which would be the subject of her only published work. But despite her insistence at the end of the letter that

> I have a <u>formidable list</u> of <u>small matters</u> down, against I see you[9]

this is the last mathematical letter to De Morgan from her that we have. The remarkable correspondence course was at an end.

43 THE closing lines of Ada's last letter of substance to De Morgan, 21 November 1841. Note the reference to the 'Numbers of Bernoulli'.

I only allude to $(x^3)^3$, instead of $(x^3)^2$ as I make it. See page 444, at the bottom; (2nd column):

"Where $B_0, B_1, \&c$ are the values of fy and its "successive diff-co's when $y = 0$, $\&c$, $\&c$"

Surely it should be when $y = 1$.

The same as when immediately afterwards, (see page 445, 1st column, at the top), in developping $(2+\Delta)^{-1} fy$; $B_0, B_1, \&c$ are the values of fy & its Co-efficients when $y = 2$, $\&c$, $\&c$. —

I have referred to Numbers of Bernoulli & to Differences of Nothing; in consequence of reading this Article Operation. And find that I must read that on Series also. —

I left off at page 165 of the Calculus; & suppose that I may now resume it, (when I return here that is). —

I will not trouble you further in this letter. But I have a formidable list of small matters down, against I see you. —

Yours most sincerely
A. A. Lovelace —

List of Operations

1. Algebraic Addition
2. ———— Subtraction
3. Alg^c Add^n with nk figures
4. ———— Subt^n with nk figures
5. Ascertaining if a variable is zero
6. ———————————————— has a + or − sign
7. Reversing the Accident^l sign of a variable
8. Stepping up or multiplying any number by 10
9. Stepping down or mult^y any number by 10
10. Count^g N^o of digits in Variable & sending it to Sto
11. Differences
12. Multiplication without Table.

THE ANALYTICAL ENGINE

Charles Babbage started work on his Analytical Engine in the mid-1830s with the idea of creating a new calculating machine that could 'eat its own tail',[1] by which he meant that it could modify its calculation while it was running. It would do this through pausing during a calculation, and using the values it had already calculated to choose between two possible next steps. Babbage listed the basic operations that such a machine, with large enough memory, would need if it was to execute 'the whole of the developments and operations of analysis',[2] in other words any calculation that could be conceived of at the time. We now know that the basic operations he described are what is needed to compute anything that can be computed by any modern computer. This means that the Analytical Engine would have been, in modern terms, a *general-purpose computer*, a concept first identified by Alan Turing in the 1930s.

The Analytical Engine was never built, but many aspects of its astonishing design were recorded in immaculate detail in Babbage's drawings and mechanical notation. It was to be programmed by means of punched cards, similar to those used in the weaving looms designed by Joseph Marie Jacquard. Separate decks of cards made up what we would now call the program,

44 Charles Babbage's list of the operations of the Analytical Engine. Alan Turing showed that any computer which can carry out these operations is, in modern terminology, a 'general purpose computer'.

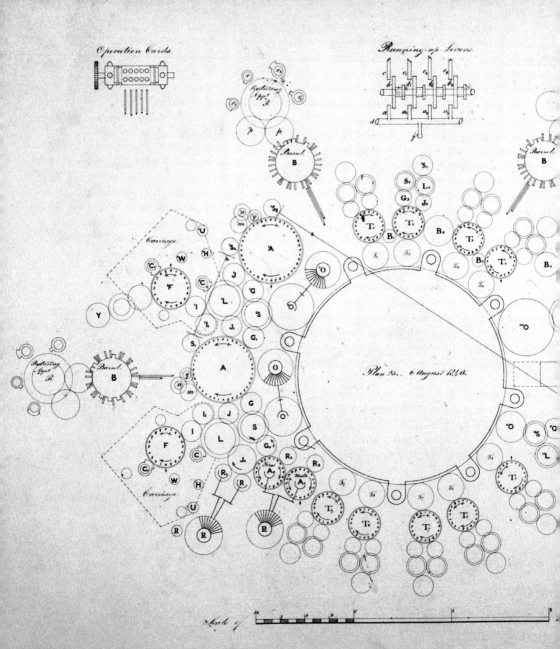

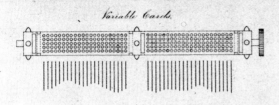
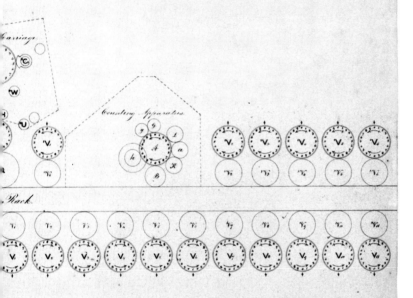
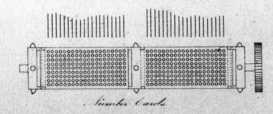
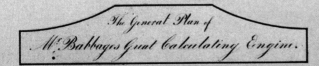

45 Charles Babbage's design drawing for the Analytical Engine.

46 'The Analytical Engine' by Sydney Padua, from *The Thrilling Adventures of Lovelace and Babbage*.

and gave the starting values for the computations. A complex mechanism allowed the machine to repeat a deck of cards, so as to execute a loop. The hardware involved many new and intricate mechanisms and was conceived on a massive scale. The central processing unit, which Babbage called the Mill, would be 15 feet tall; the memory, or Store, holding a hundred 50-digit numbers would be 20 feet long (Babbage even considered machines with ten times that capacity); and other components included a printer, card punch and graph plotter. Babbage estimated it would take three minutes to multiply two 20-digit numbers. A machine of that size would indeed have required steam power.

The Analytical Engine, had it been built to Babbage's plans in the early 1840s.

1. **The Store** (hard disk, or memory). 2. **The Mill** (Central Processing Unit). 3. **Steam Engine** (power). 4. **Printer** (printer, round the other side). 5. **Operation Cards** (the program). 6. **Variable Cards** (Addressing system) 7. **Number Cards** (for entering numbers). 8. **The Barrel Controllers** (microprograms).

Disillusioned by what he saw as lack of support from the British scientific establishment, Babbage looked for funding abroad. In 1840 Italian scientists invited him to Turin, among them Luigi Menabrea, who in October 1842 published the first account of the Engine, in French, based on Babbage's lectures. Ada Lovelace had been thinking for some time about how she might contribute to Babbage's projects. Another scientific friend, Charles Wheatstone, asked if she would translate Menabrea's article, and Babbage suggested she expand it with a number of appendices. After several months of furious effort by them both, with Lord Lovelace sometimes dragged in as a copyist, the resulting paper was published in Taylor's *Scientific Memoirs* in August 1843. It was signed only with her initials, A.A.L., and of its sixty-six pages, forty-one are her appendices.

The paper is most famous for the final appendix, 'Note G', which shows how the machine worked by calculating the Bernoulli numbers. Lovelace first explained what these numbers are. The formula

$$\frac{x}{e^x - 1}$$

is equal to an infinite series

$$\frac{x}{e^x - 1} = 1 - \frac{x}{2} + B_1 \frac{x^2}{2} + B_3 \frac{x^4}{2 \cdot 3 \cdot 4} + B_5 \frac{x^6}{2 \cdot 3 \cdot 4 \cdot 5 \cdot 6} + \ldots$$

and the values B_1, B_3, B_5 and so on are called the Bernoulli numbers. The first few are

$$B_1 = \frac{1}{6}, \quad B_3 = -\frac{1}{30}, \quad \ldots.$$

In the paper Lovelace gave a formula for the numbers:

$$0 = -\frac{1}{2} \cdot \frac{2n-1}{2n+1} + B_1\left(\frac{2n}{2}\right) + B_3\left(\frac{2n \cdot (2n-1) \cdot (2n-2)}{2 \cdot 3 \cdot 4}\right)$$
$$+ B_5\left(\frac{2n \cdot (2n-1)\ldots(2n-4)}{2 \cdot 3 \cdot 4 \cdot 5 \cdot 6}\right) + \ldots + B_{2n-1}.$$

She chose this formula from a number of possibilities because the 'object is not simplicity or facility of computation, but the illustration of the powers of the engine'.[3]

This formula really represents a whole set of formulae, one for each value of n, and Lovelace's method allows each successive Bernoulli number to be worked out, once the preceding ones are known. So if we put n equal to 1 in the formula above, we get

$$0 = -\frac{1}{2}\frac{(2-1)}{(2+1)} + B_1\frac{2 \cdot 1}{2}$$

and so

$$B_1 = \frac{1}{6}.$$

If we put n equal to 2 in the formula we get

$$0 = -\frac{1}{2}\frac{(2 \cdot 2 - 1)}{(2 \cdot 2 + 1)} + B_1\frac{2 \cdot 2}{2} + B_3$$

so that

$$B_3 = \frac{3}{10} - (B_1 \times 2)$$

and as we already know that

$$B_1 = \frac{1}{6}$$

we have

$$B_3 = \frac{3}{10} - (\frac{1}{6} \times 2) = -\frac{1}{30}.$$

In principle we can carry on in the same way and calculate all the Bernoulli numbers.

The paper contained a detailed explanation of how the various quantities involved are fetched from the Store, used in calculation in the Mill, and moved back again, according to the instructions on the cards. The process is illustrated using a large table, whose columns represent the values of the data, the variables and the intermediate results, as the engine carried out each stage of the calculation (FIGURE 47).

This table is often described as 'the first computer program', though Lovelace wrote, more accurately, that it 'presents a complete simultaneous view of all the successive changes' in the components of the machine, as the calculation progresses. In other words, the table is what computer scientists would now call an 'execution trace'. The 'program', had the idea existed at the time, would have been the deck of punched cards that caused the machine to make those successive changes. Babbage's designs were rather unclear about aspects of how the cards would be manipulated, so it is hard to reconstruct the exact program. Such tables were still used as a method for explaining computation a hundred years later, when Geoff Toothill drew a similar diagram to illustrate the working of the first stored program computer, the 'Manchester Baby'.

'Note G' is the culmination of Lovelace's paper, following many pages of detailed explanation of the operation of the Engine and the cards, and of the notation of the tables. The paper shows Lovelace's obsessive attention to mathematical details – it also shows her imagination in thinking about the bigger picture.

Lovelace observed a fundamental principle of the machine, that the operations, defined by the cards, are separate from the data and the results. She observed that the machine might act upon things other than numbers, if those things satisfied mathematical rules.

OVERLEAF
47 'Diagram for the computation by the Engine of the Numbers of Bernoulli', from Ada Lovelace's 'Note G' to her 1843 translation of C.F. Menabrea's 'Sketch of the Analytical Engine invented by Charles Babbage'. This diagram is sometimes called 'The first computer program'.

Diagram for the computation by the Engine of

Number of Operation.	Nature of Operation.	Variables acted upon.	Variables receiving results.	Indication of change in the value on any Variable.	Statement of Results.	Data. $^1V_1$ ○ 0 0 1 $\boxed{1}$	$^1V_2$ ○ 0 0 2 $\boxed{2}$	$^1V_3$ ○ 0 0 4 $\boxed{n}$	$^0V_4$ ○ 0 0 0 ☐	$^0V_5$ ○ 0 0 0 ☐	$^0V_?$
1	×	$^1V_2 \times ^1V_3$	$^1V_4, ^1V_5, ^1V_6$	$\left\{\begin{array}{l}^1V_2 = ^1V_2\\^1V_3 = ^1V_3\end{array}\right\}$	$= 2n$	...	2	n	$2n$	$2n$	2
2	−	$^1V_4 - ^1V_1$	$^2V_4$	$\left\{\begin{array}{l}^1V_4 = ^2V_4\\^1V_1 = ^1V_1\end{array}\right\}$	$= 2n-1$	1	...	...	$2n-1$		
3	+	$^1V_5 + ^1V_1$	$^2V_5$	$\left\{\begin{array}{l}^1V_5 = ^2V_5\\^1V_1 = ^1V_1\end{array}\right\}$	$= 2n+1$	1	...	...	...	$2n+1$	
4	÷	$^2V_5 \div ^2V_4$	$^1V_{11}$	$\left\{\begin{array}{l}^2V_5 = ^0V_5\\^2V_4 = ^0V_4\end{array}\right\}$	$= \dfrac{2n-1}{2n+1}$	...	...	...	0	0	
5	÷	$^1V_{11} \div ^1V_2$	$^2V_{11}$	$\left\{\begin{array}{l}^1V_{11} = ^2V_{11}\\^1V_2 = ^1V_2\end{array}\right\}$	$= \dfrac{1}{2}\cdot\dfrac{2n-1}{2n+1}$	...	2	...	...	...	
6	−	$^0V_{13} - ^2V_{11}$	$^1V_{13}$	$\left\{\begin{array}{l}^2V_{11} = ^0V_{11}\\^0V_{13} = ^1V_{13}\end{array}\right\}$	$= -\dfrac{1}{2}\cdot\dfrac{2n-1}{2n+1} = A_0$	...	...	...	...	...	
7	−	$^1V_3 - ^1V_1$	$^1V_{10}$	$\left\{\begin{array}{l}^1V_3 = ^1V_3\\^1V_1 = ^1V_1\end{array}\right\}$	$= n-1 (=3)$	1	...	n	...	...	
8	+	$^1V_2 + ^0V_7$	$^1V_7$	$\left\{\begin{array}{l}^1V_2 = ^1V_2\\^0V_7 = ^1V_7\end{array}\right\}$	$= 2+0 = 2$	...	2	...	...	...	
9	÷	$^1V_6 \div ^1V_7$	$^3V_{11}$	$\left\{\begin{array}{l}^1V_6 = ^1V_6\\^0V_{11} = ^3V_{11}\end{array}\right\}$	$= \dfrac{2n}{2} = A_1$	...	...	...	...	...	2
10	×	$^1V_{21} \times ^3V_{11}$	$^1V_{12}$	$\left\{\begin{array}{l}^1V_{21} = ^1V_{21}\\^3V_{11} = ^3V_{11}\end{array}\right\}$	$= B_1 \cdot \dfrac{2n}{2} = B_1 A_1$	...	...	...	...	...	
11	+	$^1V_{12} + ^1V_{13}$	$^2V_{13}$	$\left\{\begin{array}{l}^1V_{12} = ^0V_{12}\\^1V_{13} = ^2V_{13}\end{array}\right\}$	$= -\dfrac{1}{2}\cdot\dfrac{2n-1}{2n+1} + B_1 \cdot \dfrac{2n}{2}$	...	...	...	...	...	
12	−	$^1V_{10} - ^1V_1$	$^2V_{10}$	$\left\{\begin{array}{l}^1V_{10} = ^2V_{10}\\^1V_1 = ^1V_1\end{array}\right\}$	$= n-2 (=2)$	1					
13	−	$^1V_6 - ^1V_1$	$^2V_6$	$\left\{\begin{array}{l}^1V_6 = ^2V_6\\^1V_1 = ^1V_1\end{array}\right\}$	$= 2n-1$	1	...	...	...	...	$2n$
14	+	$^1V_1 + ^1V_7$	$^2V_7$	$\left\{\begin{array}{l}^1V_1 = ^1V_1\\^1V_7 = ^2V_7\end{array}\right\}$	$= 2+1 = 3$	1	...	...	...	...	
15	÷	$^2V_6 \div ^2V_7$	$^1V_8$	$\left\{\begin{array}{l}^2V_6 = ^2V_6\\^2V_7 = ^2V_7\end{array}\right\}$	$= \dfrac{2n-1}{3}$	...	...	...	...	...	$2n$
16	×	$^1V_8 \times ^3V_{11}$	$^4V_{11}$	$\left\{\begin{array}{l}^1V_8 = ^0V_8\\^3V_{11} = ^4V_{11}\end{array}\right\}$	$= \dfrac{2n}{2}\cdot\dfrac{2n-1}{3}$	...	...	...	...	...	
17	−	$^2V_6 - ^1V_1$	$^3V_6$	$\left\{\begin{array}{l}^2V_6 = ^3V_6\\^1V_1 = ^1V_1\end{array}\right\}$	$= 2n-2$	1	...	...	...	...	$2n$
18	+	$^1V_1 + ^2V_7$	$^3V_7$	$\left\{\begin{array}{l}^2V_7 = ^3V_7\\^1V_1 = ^1V_1\end{array}\right\}$	$= 3+1 = 4$	1	...	...	...	...	
19	÷	$^3V_6 \div ^3V_7$	$^1V_9$	$\left\{\begin{array}{l}^3V_6 = ^3V_6\\^3V_7 = ^3V_7\end{array}\right\}$	$= \dfrac{2n-2}{4}$	...	...	...	...	...	$2n$
20	×	$^1V_9 \times ^4V_{11}$	$^5V_{11}$	$\left\{\begin{array}{l}^1V_9 = ^0V_9\\^4V_{11} = ^5V_{11}\end{array}\right\}$	$= \dfrac{2n}{2}\cdot\dfrac{2n-1}{3}\cdot\dfrac{2n-2}{4} = A_3$	...	...	...	...	...	
21	×	$^1V_{22} \times ^5V_{11}$	$^0V_{12}$	$\left\{\begin{array}{l}^1V_{22} = ^1V_{22}\\^0V_{12} = ^2V_{12}\end{array}\right\}$	$= B_3 \cdot \dfrac{2n}{2}\cdot\dfrac{2n-1}{3}\cdot\dfrac{2n-2}{3} = B_3 A_3$	...	...	...	...	...	
22	+	$^2V_{12} + ^2V_{13}$	$^3V_{13}$	$\left\{\begin{array}{l}^2V_{12} = ^0V_{12}\\^2V_{13} = ^3V_{13}\end{array}\right\}$	$= A_0 + B_1 A_1 + B_3 A_3$	...	...	...	...	...	
23	−	$^2V_{10} - ^1V_1$	$^3V_{10}$	$\left\{\begin{array}{l}^2V_{10} = ^3V_{10}\\^1V_1 = ^1V_1\end{array}\right\}$	$= n-3 (=1)$	1	...	...	...	...	

Here follows a repetit

| 24 | + | $^4V_{13} + ^0V_{24}$ | $^1V_{24}$ | $\left\{\begin{array}{l}^4V_{13} = ^0V_{13}\\^0V_{24} = ^1V_{24}\end{array}\right\}$ | $= B_7$ .... | ... | ... | ... | ... | ... | |
| 25 | + | $^1V_1 + ^1V_3$ | $^1V_3$ | $\left\{\begin{array}{l}^1V_1 = ^1V_1\\^1V_3 = ^1V_3\\^5V_6 = ^0V_6\\^5V_7 = ^0V_7\end{array}\right.$ by a Variable-card. by a Variable card. | $= n+1 = 4+1 = 5$ .... | 1 | ... | $n+1$ | ... | ... | |

of Bernoulli. See Note G. (page 722 *et seq.*)

		Working Variables.				Result Variables.		
$^0V_9$ ○ 0 0 0 0 □	$^0V_{10}$ ○ 0 0 0 0 □	$^0V_{11}$ ○ 0 0 0 0 □	$^0V_{12}$ ○ 0 0 0 0 □	$^0V_{13}$ .... ○ 0 0 0 0 □	$^1V_{21}$ ○ 0 0 0 0 B_1 in a decimal fraction. □ B_1	$^1V_{22}$ ○ 0 0 0 0 B_3 in a decimal fraction. □ B_3	$^1V_{23}$ ○ 0 0 0 0 B_5 in a decimal fraction. □ B_5	$^0V_{24}$ ... ○ 0 0 0 0 □ B_7
...	...	$\dfrac{2n-1}{2n+1}$						
...	...	$\dfrac{1}{2}\cdot\dfrac{2n-1}{2n+1}$						
...	...	0	....	$-\dfrac{1}{2}\cdot\dfrac{2n-1}{2n+1} = A_0$				
...	$n-1$							
...	...	$\dfrac{2n}{2} = A_1$						
...	...	$\dfrac{2n}{2} = A_1$	$B_1\cdot\dfrac{2n}{2}=B_1 A_1$	....	B_1			
...	...	....	0	$\left\{-\dfrac{1}{2}\cdot\dfrac{2n-1}{2n+1}+B_1\dfrac{2n}{2}\right\}$				
...	$n-2$							
...	...	$\dfrac{2n}{2}\cdot\dfrac{2n-1}{3}$						
$\dfrac{2n-2}{4}$ 0	...	$\left\{\dfrac{2n}{2}\cdot\dfrac{2n-1}{3}\cdot\dfrac{2n-2}{3} = A_3\right\}$						
...	...	0	$B_3 A_3$	....	....	B_3		
...	...	....	0	$\left\{A_3 + B_1 A_1 + B_3 A_3\right\}$				
...	$n-3$							

thirteen to twenty-three.

| ... | ... | .... | .... | .... | .... | .... | .... | B_7 |

> Supposing that the fundamental relations of pitched sounds in the science of harmony and of musical composition were susceptible of such expression and adaptations, the engine might compose elaborate and scientific pieces of music of any degree of complexity or extent.[4]

She thought about how the engine might do algebra, how it *'weaves algebraical patterns* just as the Jacquard loom weaves flowers and leaves' and how it might make new discoveries:

> We might even invent laws for series or formulæ in an arbitrary manner, and set the engine to work upon them, and thus deduce numerical results which we might not otherwise have thought of obtaining.[5]

These led her to think about what we now call artificial intelligence, though she argued that the engine is not capable of original ideas:

> The Analytical Engine has no pretensions whatever to *originate* anything. It can do whatever we *know how to order it* to perform.[6]

Alan Turing disagreed. In a famous paper on 'Computing machinery and intelligence' he challenged what he called 'Lady Lovelace's objection'. He suggested that the machine could be 'ordered' to be original, by programming it to produce unpredictable answers.

Lovelace's thoughts about using the machine are very familiar to present-day programmers. She understood how complicated programming is, and how difficult it can be to get things right, as

> There are frequently several distinct sets of effects going on simultaneously; all in a manner independent of each other, and yet to a greater or less degree exercising a mutual influence.[7]

25th August, 1843

$$\frac{x}{x + \frac{x^2}{2} + \frac{x^3}{2\cdot 3} + \cdots}$$

$$\frac{1}{1 + \frac{x}{2} + \cdots}$$

$$\frac{0}{0} \qquad \frac{\varphi x}{\psi x} \qquad \frac{\varphi' x}{\psi' x}$$

$$\frac{x}{\varepsilon^x - 1} \qquad \frac{1}{\varepsilon^x}$$

$$\text{Co. of } x^{2n} \text{ in } \frac{\frac{1}{2}x}{\varepsilon^{\frac{x}{2}} + 1}$$

$$= \frac{1}{2^{2n}} \text{ co. of } x^{2n} \text{ in } \frac{x}{\varepsilon^x + 1}$$

$$= \frac{1}{2^{2n}} \text{ Co. of } x^{2n-1} \text{ in } \frac{1}{\varepsilon^x + 1}$$

$$= \frac{1}{2^{2n}} \cdot \frac{1}{1.2.3\ldots 2n-1} \left(\frac{d}{dx}\right)^{2n-1} \frac{1}{\varepsilon^x + 1}$$

when $x = 0$

48 Ada Lovelace, note with a formula for the Bernoulli numbers.

Also the rest of Note D. There is still one trifling misapprehension about the Va-riable cards. A Variable card may order any number of Variables to receive the same number upon them at the same instant of time. But a Variable card never can be directed to order more than one Variable to be given off at once because the mill could not receive it and the mechanism would not permit it. All this it was impossible for you to know by intuition and the more I read your notes the more surprised I am at them and regret not having earlier explored so rich a vein of the noblest metal.

The account of them stands thus

A Sent to Lady L.
B With C.B.
C Ditto
D Sent to Lady L
E With C.B

F Returned by Lady L
G Where is it gone??
H With C.B

49 Letter from Charles Babbage to Ada Lovelace, 2 July 1843, as they work on the manuscript of the paper.

...not seen Mr Wheatstone and
...ed to write untill I can pos[s]-
... The whole of the notes in his hands.
...will attend your commands
— and am
Ever most truly Yours
C Babbage

...March 9
...y 1843

50 Letter from Ada Lovelace to Charles Babbage, 10 July 1843, suggesting the paper include the Bernoulli numbers as 'an example of how an implicit function may be worked out by the engine, without having been worked out by human head & hands first'.

...ll send down to the
... before tomorrow
...ning, Brookes's Formulae,
...also the Reports of the
...yal Society on your
...achine. I suppose you
...n get it easily, & I
...rticularly want to see
..., before I see you on
...ed'y Morn'g. —

It appears to me
...t I am working up the
...tes with much success;
...that even if the book
...e delayed in its
...publication, a week or

two, in consequence, it would
be worth Mr Taylor's while
to wait. I will have it
well & fully done; or not
at all.

I want to put in
something about Bernoulli's
Numbers, in one of my
Notes, as an example of
how an implicit function
may be worked out by
the engine, without having
been worked out by
human head & hands
first. Give me the necessary
data & formulae.

Yours ever
A. A. L.

And, echoing a concern of every programmer ever, she also appreciated the need to

> reduce to a minimum the time necessary for completing the calculation.[8]

Lovelace's paper is an extraordinary accomplishment, probably understood and recognized by very few in its time, yet still perfectly understandable nearly two centuries later. It covers algebra, mathematics, logic and even philosophy; a presentation of the unchanging principles of the general-purpose computer; a comprehensive and detailed account of the so-called 'first computer program'; and an overview of the practical engineering of data, cards, memory and programming.

Lovelace and Babbage's collaboration by letter, as they exchanged versions of the table for the Bernoulli numbers, echoes the frustrations of all collaborators – 'Where is it gone?'[9] wrote Babbage as they lost track of 'Note G' (FIGURE 49). Towards the end of the work tempers became frayed. Lovelace refused to let Babbage add to the paper a strong criticism of the British government; and Babbage turned down her offer to become further involved in organizing the building of the engine.

However, Babbage continued to speak admiringly of her, writing to Michael Faraday of

> that Enchantress who has thrown her magical spell around the most abstract of Sciences and has grasped it with a force which few masculine intellects (in our own country at least) could have exerted over it.[10]

They did not collaborate again, but remained friends: Lovelace's letters to Babbage are full of details of the mathematics books she is reading, the progress of her children, and the antics of her dogs, chickens and starlings. In the last year of her life, Babbage accompanied the now frail Ada to the Great Exhibition, and encouraged her to 'put on worsted stockings, cork soles and every other thing which can keep you warm'.[11] To his annoyance, none of his machines were displayed there.

51 From *Dickinson's Comprehensive Pictures of the Great Exhibition*, 1851. Babbage went with Ada, but none of his machines were displayed.

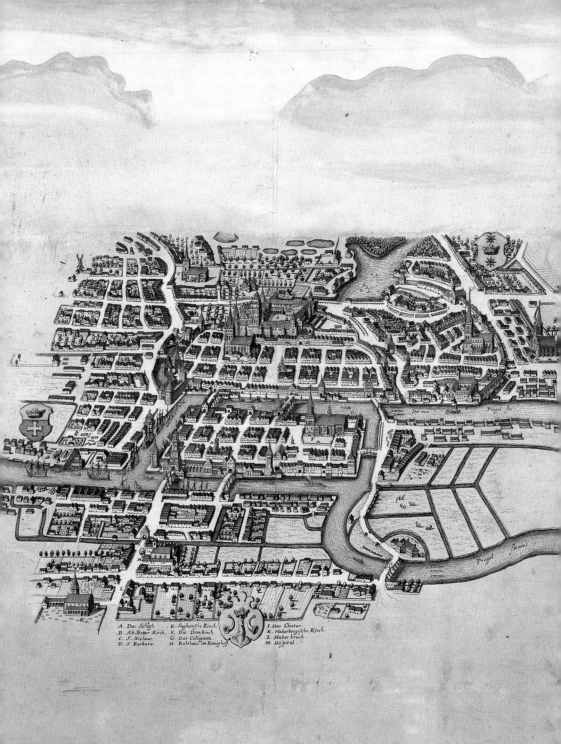

8

MATHEMATICAL PUZZLES AND SPECULATIONS

Mathematicians usually throw away their scrap paper – but the scribbled manuscript shown overleaf (FIGURE 53) has survived in the archives. One can imagine the animated conversation that produced it, Charles Babbage scrawling with a quill pen that very much needs sharpening, and Ada Lovelace leaning across sideways to add details in pencil.

The most striking part illustrates a famous question in mathematics called the Königsberg Bridge Problem. Engravings of the ancient city of Königsberg, then in Eastern Prussia, now Kaliningrad in Russia, show that the four regions of the city were separated by the River Pregel, and were joined by seven bridges. It is said that the Swiss mathematician Leonhard Euler watched people playing a popular game of trying to walk around the city, crossing each bridge exactly once before returning to their starting point. The problem was that no one seemed able to do it, which led people to wonder whether it was possible. Euler was surprised that people thought of this problem as mathematics, which to him meant algebra or geometry, and he wrote in 1736:

> I do not understand why you expect a mathematician to produce it, rather than any one else, for the solution is based

52 Map showing the seven bridges of Königsberg across the River Pregel, by Matthias Merian, 1641.

$4 \times 3 + 4 \times 2 + 1 \times 4$
$12 + 8 + 4 = 24$

1 No return

2 No ending except
4   in island

3 there are an odd N ^{o} of Bridges
   you do not begin you must end
   it

3 around 3 4 6 2
              2 4 6 8

5 = 4

PREVIOUS PAGES
53 Undated manuscript of mathematical puzzles, c.1840, in the hands of Ada Lovelace and Charles Babbage. The main part of the document is a description in pictures and words of the Königsberg Bridge Problem. It also includes a magic square (lower centre, rotated) and a diagram of Pythagoras' theorem (mid-centre, rotated).

on reason alone, and its discovery does not depend on any mathematical principle.[1]

Euler solved the problem by showing that such a path could only be found if each region of land had an even number of bridges leaving it.

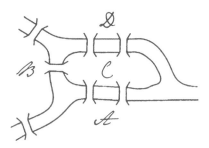

Today, we would draw the map of Königsberg as a graph whose nodes represent land areas and whose edges represent bridges.

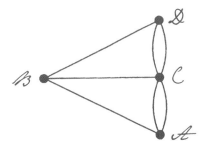

From this graph, we can see that the number of edges (or bridges) coming out of each node (or land region) is 3, 3, 5 and 3. Since these are all odd numbers, the citizens of Königsberg were doomed to fail – a path which crosses every bridge exactly once is impossible.

In the document, Babbage has drawn several alternative patterns of islands and bridges. Someone, presumably Ada Lovelace, has attempted to trace a path on one of them, and if you look closely you can see where the pencil has dragged through the wet ink. Babbage also provided a few scribbled instructions: 'No returns', 'If there are an odd Nº of Bridges, if you do not begin you must end in it', 'No ending except in island'. These instructions do more than tell you if there is a solution: they form a set of rules – an algorithm – for solving the general problem. The document would be even more exciting if this was the first known description of the algorithm, but it is not. We know that the French mathematician Louis Poinsot described it in a book published in 1810, and we even know that Babbage owned the book.

Euler's account of the Königsberg Bridge Problem is one of the earliest examples of the kind of mathematics now called graph theory. Graphs can model all kinds of networks and communication structures such as roads, computer circuitry or connections in social networks. Algorithms to find different kinds of paths through graphs are now used by satnavs to work out directions, in biology to reconstruct DNA sequences from fragments, and in social networks to plan marketing campaigns.

In 1834 Ada Byron had puzzled her tutor Dr King with a question related to Pythagoras' theorem. The diagram of squares and triangles shows a proof of Pythagoras' theorem which only works when the sides of the triangle are 1, 2 and $\sqrt{5}$ units long. To show this, cut up the square on the side of length 2 units as shown, and rearrange the pieces in the square on the hypotenuse. You are left with a small square in the middle, of size 1 unit, which exactly matches the square on the short side of the triangle, proving the result.

The other jottings and doodles on the page are as yet undeciphered, though the dots and lines, so confidently drawn in pencil to the bottom left of the image, seem to be capturing some kind of algorithm. And the whole document is remarkable, capturing a lively informal conversation about algorithms by two people who probably understood more about computing than anyone else in the world at the time.

The manuscript also shows a magic square, a block of numbers arranged so that their sum in any direction is always the same. A diagram near the magic square shows an algorithm for constructing it, based on writing the numbers 1, ..., 9 in a grid and swapping them round.

$$\begin{matrix} 2 & 9 & 4 \\ 7 & 5 & 3 \\ 6 & 1 & 8 \end{matrix}$$

Identical diagrams of a magic square and grid appear among some notes Babbage wrote while he was staying with the Lovelaces in September 1848. The notes explain part of his design of a machine to play noughts and crosses (which he called 'Tit-tat-to'). He worked on this for some twenty years, inventing a way to represent board positions, and studying the strategy for the game and how to build a machine that would play it. At one stage he hoped, somewhat unrealistically, to raise money by charging a hefty admission fee for demonstrations. He was encouraged by Ada Lovelace, who wrote to him after his visit:

> You say nothing of Tit-tat-to – in yr last. I am alarmed lest it should never be accomplished. I want you to complete something; especially if the something is likely to produce silver & golden somethings.[2]

Babbage thought about chess in a similar way and his ideas applied to any game of strategy. A few years earlier Ada had written to him about peg solitaire, a game in which a player follows certain rules to remove pegs or marbles one by one from a board, leaving just one behind (FIGURE 54).

> I want to know if the problem admits of being put into a mathematical formula, & solved in this manner. I am convinced myself that it does, though I cannot do it. There must be a definite principle, a compound I imagine of numerical and geometrical properties, on which the solution depends, & which can be put into symbolic language.[3]

Ada's idea was many years ahead of its time, and we now know that there is indeed such a 'definite principle'.

Ada also wrote about mathematics and discovery, and suggested to Babbage that she might work out

> a system, of the principles and methods of discovery, elucidating the same with examples. I am already noting down a list of discoveries hitherto made, in order myself to examine into their history, origin and progress. One first & main point, whenever and wherever I introduce the subject, will be to define and classify all that is to be included under the term discovery.[4]

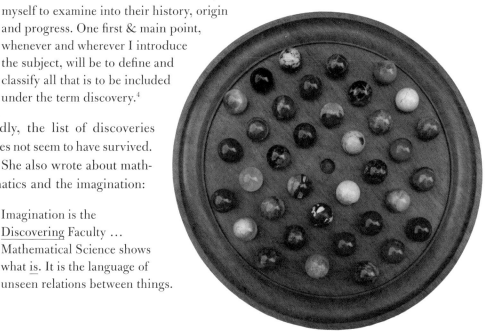

54 A board for playing peg solitaire: Ada Lovelace speculated that there should be a mathematical way to solve the puzzle.

Sadly, the list of discoveries does not seem to have survived.

She also wrote about mathematics and the imagination:

> Imagination is the Discovering Faculty ... Mathematical Science shows what is. It is the language of unseen relations between things.

55 An undated fragment of a letter from Ada Lovelace, probably to her mother, in which she discusses 'poetical science'.

But to use & apply that language we must be able fully to appreciate, to feel, to seize, the unseen, the unconscious. Imagination too shows what <u>is</u>, the <u>is</u> that is beyond the senses. Hence she is ... especially cultivated by the truly Scientific, those who wish to enter into the worlds around us![5]

The sentiment is reminiscent of the poet Coleridge, and Lovelace herself considered writing mathematical poetry. This, she told her husband, would surprise her mother, as it would be 'poetry of an <u>unique</u> kind; far more <u>philosophical</u> & higher in it's nature than aught the world has perhaps yet seen', which would go beyond 'a mathematical astronomical view of the heavens' so that if she were to write verses on the Moon 'the subject would be the living things of our Satellite ... and the aspects our planet presents to the spirits of the Moon ... a very sublime poem, but not a word therein of mathematics & the laws of motion'.[6]

She wrote to her mother:

You will not concede me <u>philosophical</u> <u>poetry</u>. Invert the order! Will you give me <u>poetical</u> <u>philosophy</u>, <u>poetical</u> <u>science</u>?[7]

This remark has become famous for capturing the breadth of Lovelace's approach to mathematics (FIGURE 55).

person not yet 30, (& with all my sufferings, I am yet vigorous enough I am sure), may do anything, if they will but go to school. It is no bad epoch of life to begin education from.

Are we d'accord now do you think, about poetry & music & philosophy? — I don't consider that as yet I have made anything like full use of music, and often much foolish abuse indeed.

You will not concede me philosophical poetry. I invert the order! Will you give me poetical philosophy, poetical science? —

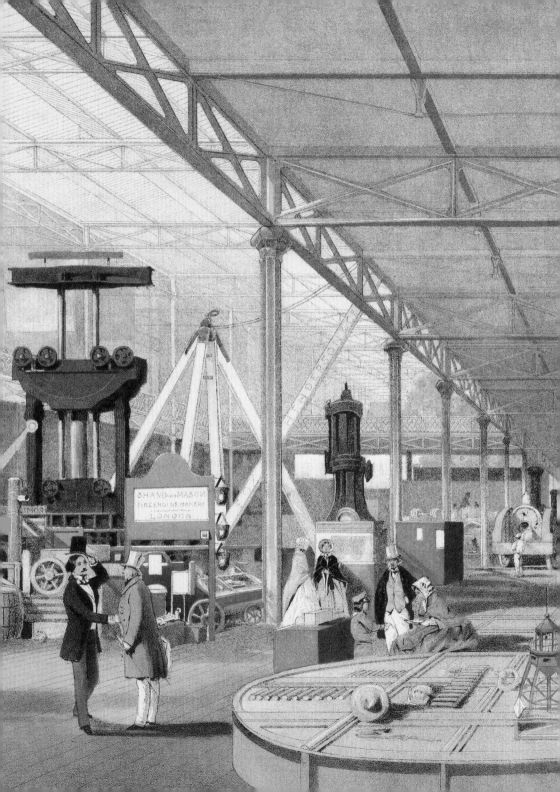

9

THE FINAL YEARS

Ada Lovelace maintained an extraordinary number of scientific interests until the end of her life. Her correspondence refers to agriculture, algebra, arithmetic, astronomy, chemistry, electricity, geometry, magnetism, mathematics, mesmerism, natural science, phrenology, photography, physiology, statistics and the telegraph.

She was not alone. The word 'scientist' was first used in print by Babbage's friend William Whewell in an 1834 review of a book by Mary Somerville. Scientific publications could be bestsellers, for example *Vestiges of the Natural History of Creation*, published anonymously in 1844, which presented controversial ideas hinting at Darwin's theory of evolution. Prince Albert, Charles Darwin and Ada Lovelace were among those suspected of being the author, whose true identity, Robert Chambers, was not revealed for forty years.

There was an increasing interest in gathering, publishing and analysing data, to make discoveries of practical use in fields like navigation, insurance and agriculture, promoted by new organizations like the Royal Statistical Society. Ada's husband, the Earl of Lovelace, was an early member. He was a reform-minded landowner, and published a number of articles on agriculture, such as 'Method of Growing Beans and Cabbages on the Same

56 The Machine Room at the Great Exhibition of 1851, which included printing presses, farm machinery, railway engines, and even a steam-powered demonstration of the spinning of cotton. Queen Victoria spent two hours going round it, finding it 'excessively interesting and instructive ... what used to be done by hand and take months doing is now accomplished in a few minutes by the most beautiful machinery'.

Ground'. His 1848 article 'On Climate in Connection with Husbandry' reviewed the work of the French author Agénor de Gasparin on different mathematical theories linking climate and the yield of crops. His wife added two lengthy footnotes, marked with her initials A.A.L. She writes that

> No universally accurate law on the relations between climate and the development of plants can be obtained until a far greater range and mass of facts have been accumulated[1]

and suggests that Gasparin's theory of plant growth should be replaced with an alternative mathematical theory due to the Belgian statistician Adolphe Quetelet. Modern scientists agree with her.

Lovelace was ahead of her time in understanding the importance of photography for science, writing

> we believe that it is as yet quite unsuspected how important a part photography is to play in the advancement of human knowledge.[2]

Her footnote suggested studying the link between sunshine and plant growth with the aid of John Herschel's actinograph. This clockwork device wound a strip of photographic paper past a lens, creating an image on the paper when the sun was shining. As Lovelace pointed out, the device saved the effort of writing down regular observations on a chart.

She proposed using photography for what we would now call crowdsourcing:

> If amateurs, of either sex, would amuse their idle hours with experimenting on this subject, & would keep an accurate journal of their daily observations, we should in a few years have a mass of registered facts to compare with the observation of the Scientific.[3]

She became fascinated by mesmerism, a popular kind of hypnosis, and wrote about doing experiments to understand its effect on the chemistry of the human body. She even read a book on poisons, which she found unsatisfactory as it gave many examples but no abstract principles allowing 'deductions to be made'. She was far ahead of her time in her ambition to understand the workings of the mind through developing a 'calculus of the nervous system',[4] by

> getting cerebral phenomena such that I can put them into mathematical equations; in short a law, or laws, for the mutual actions of the molecules of brain (equivalent to the law of gravitation for the planetary & siderial [sic] world).[5]

Perhaps she imagined something like Babbage's mechanical notation, which is similar to the visual languages now used in computer chip design, and which can indeed be used to represent the interaction of molecules.

Throughout her life Ada Lovelace suffered from ill health, and was eventually diagnosed with uterine cancer. She was prescribed increasing quantities of narcotics, accompanied by

57 Ada Lovelace's letter to Woronzow Greig about her hopes for a 'calculus of the nervous system', 15 November 1844.

THE FINAL YEARS 101

alcohol to counter the side effects. This may explain the wild language of some of her private letters, particularly to her mother, where she made ambitious claims about her mathematical talent. She hoped, eventually, to leave a mathematical legacy to the world, but we do not know what she intended that legacy to be. Was she thinking of textbooks and translations, like her friend Mary Somerville; or mathematical papers in her own right, like her teacher De Morgan; or unpublished but far-reaching innovations, like Charles Babbage's mechanical notation; or broader reflections, like Chambers' *Vestiges*; or even some form of 'mathematical poetry'?

The mathematical work that she did produce shows that she was tenacious over details, determined to work things out from first principles and drawn to big questions. Augustus De Morgan thought highly of her, and believed she was capable in time of much more than the article about the Analytical Engine, writing to Lady Byron of her

> power of thinking ... utterly out of the common way for any beginner, man or woman[6]

and observing that,

> Had any young [male] beginner, about to go to Cambridge, shewn the same power[s], I should have prophesied ... that they would have certainly made him an original mathematical investigator, perhaps of first-rate eminence.[7]

But he also expressed concern that mathematics might weaken her daughter's fragile health.

Success for a woman in science in nineteenth-century Britain required a combination of many fortunate circumstances: access to education and books; talent and ambition, recognized and nurtured by herself and others; good health; support from husband and parents; and wealth and social standing. Women were unable to attend university, or to join scientific societies or to access scholarly libraries. A woman needed a competent governess and tutors, and then, as her interests matured, mentors to give her access to scientific meetings and papers, and to work with her as intellectual equals, offering criticism as well as flattery. She needed to reconcile her talent and ambition with widespread concerns, among women and men, that mathematics was not an appropriate or decorous activity for women, that women were incapable of serious mathematical work or that they were not strong enough to undertake it.

Ada Lovelace died on 27 November 1852, aged just thirty-six. In the last few weeks of her life, although in great pain, she had her portrait painted, and asked for her friends to be given copies of it after her death. She was buried, at her request, beside her father at Hucknall in Nottinghamshire, and was commemorated by a monument inscribed with a poem she had written, entitled 'The Rainbow', the subject of her letter to William Frend some twenty years before.

It could be argued that Ada Lovelace achieved little of lasting mathematical significance – she was responsible for no major mathematical or scientific discovery. Indeed, when compared to many of the leading European mathematicians of the time, the same could also be said of her mentors, Babbage, De Morgan and Somerville, despite their many published works. Yet, as the letters and manuscripts in this book illustrate, her mathematical insight and understanding were almost unique for a woman of the period. So while we might regret that she did not do more, we can also celebrate what she

58 In the last few months of her life Ada Lovelace, in great pain, had her portrait painted by Henry Wyndham Phillips. This daguerreotype copy was made, at her request, to give to friends after her death.

did achieve, and reflect on the many other talented women who had very little opportunity to achieve anything at all.

Ada Lovelace's writings, and her life, have captured the imagination of many, and her name lives on, in events celebrating women in science, and in books, plays and graphic novels.

59 Lady Byron's memorial to her daughter in Kirkby Mallory churchyard, Leicestershire, inscribed with Ada Lovelace's poem 'The Rainbow', which ends with the words 'A hidden light may burn that never dies / But bursts thro' clouds in purest hues exprest!'

NOTES

ONE
1. George Gordon, Lord Byron, *Childe Harold's Pilgrimage*, canto 3, stanza 1.
2. George Gordon, Lord Byron, Letter to Lady Melbourne, 18 October 1812, in *Famous in My Time: Byron's Letters and Journals*, Volume II: *1810–1812*, ed. Leslie A. Marchand, John Murray, London, 1973, p. 231.
3. George Gordon, Lord Byron, *Don Juan*, canto 1, stanza 12.
4. Ibid., stanza 13.
5. Samuel Taylor Coleridge, *Coleridge's Notebooks: A Selection*, ed. Seamus Perry, Oxford University Press, Oxford, 2002, p. 39.

TWO
1. Lady Byron to William Frend, 23 August 1818 (Dep. Lovelace Byron 71, fol. 35v).
2. Governess's notebook (Dep. Lovelace Byron 118/5, fol. 4r).
3. Ibid., fol. 2r.
4. Notebook kept by Lady Byron for Ada, Jun–Oct 1821 (Dep. Lovelace Byron 118/6, fols 3r–3v).
5. Governess's notebook (Dep. Lovelace Byron 118/5).
6. Ibid., fols 6v–7r.
7. Notebook kept by Lady Byron for Ada, Jun–Oct 1821 (Dep. Lovelace Byron 118/6, fol. 4r).
8. Ibid., fol. 3r.
9. Ibid., fol. 27r.
10. Ibid., fol. 28r.
11. Ibid., fols 27r–27v.
12. Ibid., fol. 27v.
13. Ada Byron to Lady Byron, 22 November 1828 (Dep. Lovelace Byron 41, fol. 78r).

14. Childhood arithmetic exercises, 1829 (Dep. Lovelace Byron 175, fol. 176r).
15. Lady Byron, Letter to Augusta Leigh, December 1823, in *The Works of Lord Byron: Letters and Journals*, vol. VI, ed. Rowland E. Prothero, John Murray, London, 1904, p. 330.
16. Ada Byron to Lady Byron, 7 April 1828 (Dep. Lovelace Byron 41, fol. 57v).
17. Ibid.
18. Ibid., fols 57v–58r.
19. Ada Byron to Lady Byron, 7 April 1828 (Dep. Lovelace Byron 41, fol. 58r).

THREE
1. Lady Byron to Dr King, describing a plan for her daughter's education (Dep. Lovelace Byron 77, fol. 35r).
2. Ibid., fol. 35v.
3. William Frend to Lady Byron (Dep. Lovelace Byron 71, fol. 153r).
4. Ada Byron to William Frend (Dep. Lovelace Byron 171, fols 127r–127v).
5. Ibid., fol. 127v.
6. Ada Byron to Dr King (Dep. Lovelace Byron 172, fol. 126v).
7. Ada Byron to Dr King (Dep. Lovelace Byron 172, fol. 131r).
8. Ada Byron to Dr King (Dep. Lovelace Byron 172, fol. 132r).
9. Dr King to Ada Byron (Dep. Lovelace Byron 172, fol. 133r).
10. William Frend to Ada Byron (Dep. Lovelace Byron 71, fol. 194r).
11. Lord William King was no relation to Ada's erstwhile tutor Dr William King.
12. Ada King to Mary Somerville (Somerville Papers, Dep. c.367, Folder MSBY-3, fols 55v–56r).
13. Ada King to Mary Somerville (Somerville Papers, Dep. c.367, Folder MSBY-3, fol. 70).
14. Mary Somerville to Ada Byron (Dep. Lovelace Byron 174, fol. 19r).
15. Ada Lovelace to Lady Byron (Dep. Lovelace Byron 41, fol. 103r).
16. Lady Byron to Ada Lovelace (Dep. Lovelace Byron 337, 11 Feb. 1838).

FOUR
1. Charles Babbage, *History of the Invention of the Calculating Engines*, Buxton Papers, Museum of the History of Science, Oxford, p. 10.
2. Charles Babbage, *Passages from the Life of a Philosopher*, Longmans, Green, London, 1864, p. 17.
3. Lady Byron to Dr King, describing the Difference Engine as 'the thinking machine', 1833 (Dep. Lovelace Byron 77, fol. 217r).
4. *Memoir of Augustus De Morgan, by his wife Sophia Elizabeth De Morgan, with selections from his letters*, Longmans, Green, London, 1882, p. 89.
5. Ada Lovelace to the Earl of Lovelace, 29 November 1844 (Dep. Lovelace Byron 166, fol. 163v).

FIVE
1. Ada Lovelace to Lady Byron, 29 July 1840 (Dep. Lovelace Byron 41, fol. 179v).
2. Ada Lovelace to Augustus De Morgan, n.d. (Dep. Lovelace Byron 170, fol. 51v).

3. Augustus De Morgan to Ada Lovelace, n.d. (Dep. Lovelace Byron 170, fol. 35r).
4. Ada Lovelace to Augustus De Morgan, n.d. (Dep. Lovelace Byron 170, fol. 83r).
5. Ada Lovelace to Lady Byron, 29 July 1840 (Dep. Lovelace Byron 41, fol. 179v).
6. Ada Lovelace to Augustus De Morgan, [January 1841] (Dep. Lovelace Byron 170, fols 91v-91r).
7. Ada Lovelace to Augustus De Morgan, 27 November [1840] (Dep. Lovelace Byron 170, fol. 149r).
8. Ada Lovelace to Augustus De Morgan, 27 November [1840] (Dep. Lovelace Byron 170, fol. 149v).
9. Ada Lovelace to Augustus De Morgan, [1840] (Dep. Lovelace Byron 170, fol. 11r).
10. Ibid.
11. Ibid.
12. Ada Lovelace to Augustus De Morgan, 13 September [1840] (Dep. Lovelace Byron 170, fols 48v-49v).
13. Augustus De Morgan to Ada Lovelace, 15 September 1840 (Dep. Lovelace Byron 170, fol. 14r).
14. Ada Lovelace to Lady Byron, 21 November [1840] (Dep. Lovelace Byron 41, fol. 187r).
15. Ada Lovelace to Augustus De Morgan, 22 December [1840] (Dep. Lovelace Byron 170, fols 70r-70v).
16. Ada Lovelace to Lady Byron, 11 January [1841] (Dep. Lovelace Byron 42, fol. 9v).

SIX

1. Ada Lovelace to Augustus De Morgan, 19 Sept. [1841] (Dep. Lovelace Byron 170, fol. 127r).
2. Charles Babbage, *Passages from the Life of a Philosopher*, Longmans, Green, London, 1864, p. 136.
3. Ada Lovelace to Augustus De Morgan, 6 Feb. [1841] (Dep. Lovelace Byron 170, fols 98r-98v).
4. Ada Lovelace to Augustus De Morgan, 10 Nov. [1840] (Dep. Lovelace Byron 170, fol. 65v).
5. Ada Lovelace to Augustus De Morgan, 19 Sept. [1841] (Dep. Lovelace Byron 170, fol. 128r).
6. Augustus De Morgan to Ada Lovelace, 15 Oct. [1840] (Dep. Lovelace Byron 170, fol. 19v).
7. Augustus De Morgan, *The Differential and Integral Calculus*, Baldwin & Cradock, London, 1842, p. vii.
8. Ibid., p. vii.
9. Ada Lovelace to Augustus De Morgan, 21 Nov. [1841] (Dep. Lovelace Byron 170, fol. 143).

SEVEN

1. Quoted in Doron Swade, *The Cogwheel Brain*, Little, Brown, London, 2000, p. 95.

2. Charles Babbage, *Passages from the Life of a Philosopher*, Longmans, Green, London, 1864, p. 136.
3. Ada Lovelace, 'Sketch of the Analytical Engine invented by Charles Babbage, Esq., by L. F. Menabrea ... with Notes by the Translator', *Scientific Memoirs*, vol. 3, ed. Richard Taylor, 1843, pp. 666-731, on p. 725.
4. Ibid., p. 694.
5. Ibid., p. 722.
6. Ibid.
7. Ibid., p. 710.
8. Ibid.
9. Charles Babbage to Ada Lovelace (Dep. Lovelace Byron 168, fol. 45v).
10. Michael Faraday, *The Correspondence of Michael Faraday*, vol. 3, ed. Frank A.J.L. James, Institution of Engineering and Technology, London, 1996, p.164.
11. Charles Babbage to Ada Lovelace (Dep. Lovelace Byron 168, fols 64r-64v).

EIGHT

1. B. Hopkins and R.J. Wilson, 'The Truth about Königsberg', *College Mathematics Journal* 35, 2004, pp. 198-207.
2. Ada Lovelace to Charles Babbage, 18 October 1848 (British Library Add MS. 37194, fol. 197r, v) .
3. Ada Lovelace to Charles Babbage, 16 February 1840 (British Library Add MS. 37191, fol. 332v) .
4. Ada Lovelace to Charles Babbage, 2 July 1843 (British Library Add. MS. 37192, fol. 335v) .
5. Ada Lovelace, Note dated 5 January 1841 (Dep. Lovelace Byron 175, fols 199a-v and 200r).
6. Ada Lovelace to the Earl of Lovelace, n.d. (Dep. Lovelace Byron 166, fols 196r-196v).
7. Ada Lovelace, probably to Lady Byron, n.d. (Dep. Lovelace Byron 44, fol. 210r).

NINE

1. Earl of Lovelace, 'On Climate in Connection with Husbandry, with reference to a work entitled "Cours d'Agriculture, par le Comte de Gasparin"', *Journal of The Royal Agricultural Society* 9, 1848, pp. 311-40.
2. Review of translation of Reichenbach, about 1846 (Dep. Lovelace Byron 175, fol. 211).
3. Ibid.
4. Ada Lovelace to Woronzow Greig, 15 November 1844 (Oxford, Bodleian Library, Mary Somerville Papers, MS. Dep. c. 367, Folder MSBY-9, fol. 255r).
5. Ibid., fol. 254r.
6. Augustus De Morgan to Lady Byron, 21 Jan. 1844 (Dep. Lovelace Byron 339).
7. Ibid.

FURTHER READING

BIOGRAPHIES OF ADA LOVELACE

Joan Baum, *The Calculating Passion of Ada Byron*, Archon Books, Hamden CT, 1986.
James Essinger, *Ada's Algorithm: How Lord Byron's Daughter Ada Lovelace Launched the Digital Age*, Gibson Square, London, 2013.
Doris Langley Moore, *Ada, Countess of Lovelace*, John Murray, London, 1977.
Miranda Seymour, *In Byron's Wake*, Simon & Schuster, London, 2018.
Dorothy Stein, *Ada: A Life and a Legacy*, MIT Press, Cambridge MA, 1985.
Betty Alexandra Toole, *Ada, the Enchantress of Numbers*, Strawberry Press, Mill Valley CA, 1992.
Benjamin Woolley, *The Bride of Science: Romance, Reason, and Byron's Daughter*, Macmillan, London, 1999.

OTHER BACKGROUND READING

Allan Chapman, *Mary Somerville and the World of Science*, Canopus, Bristol, 2004.
Raymond Flood, Adrian C. Rice and Robin J. Wilson, *Mathematics in Victorian Britain*, Oxford University Press, Oxford, 2011.
Sydney Padua, *The Thrilling Adventures of Lovelace and Babbage: The (Mostly) True Story of the First Computer,* Particular Books, London, 2015.
Laura J. Snyder, *The Philosophical Breakfast Club: Four Remarkable Friends Who Transformed Science and Changed the World*, Broadway Books, New York, 2012.
Doron Swade, *The Cogwheel Brain: Charles Babbage and the Quest to Build the First Computer*, Little Brown, London, 2000.
Robin Wilson, *Four Colors Suffice: How the Map Problem Was Solved*, Princeton University Press, Princeton NJ, 2013.

IMAGE CREDITS

The Ada Lovelace Papers reprinted by permission of Pollinger Ltd, www.pollingerltd.com

1. © UK Government Art Collection
2. © National Portrait Gallery, London
3. National Library of Scotland, John Murray Archive
4. Oxford, Bodleian Library, 1707 d.84
5. © Victoria and Albert Museum, London
6. Wellcome Library, London
7. © National Portrait Gallery, London
8. The Principal and Fellows of Somerville College, Oxford
9. The Principal and Fellows of Somerville College, Oxford
10. © Victoria and Albert Museum, London, Given by John Sheepshanks, 1857
11. Oxford, Bodleian Library, Dep. Lovelace Byron 41, fols 3v–4r
12. Oxford, Bodleian Library, Opie H 195
13. Oxford, Bodleian Library, John Johnson Collection: Games 16 (3)
14. Oxford, Bodleian Library, Dep. Lovelace Byron 175, fols 175v–176r
15. Oxford, Bodleian Library, Dep. Lovelace Byron 41, fols 57v–56r
16. Oxford, Bodleian Library, Radcliffe Science Library, *Mechanics' Magazine*, 1852 Sept 25, vol 57, Per. 1861 e. 73
17. © Victoria and Albert Museum, London, given by Isobel Constable
18. Oxford, Bodleian Library, Douce BB 698, Pl. XV
19. Oxford, Bodleian Library, Dep. Lovelace Byron 171, fol. 127
20. Oxford, Bodleian Library, Dep. Lovelace Byron 170, fols 177v–176r
21. The Principal and Fellows of Somerville College, Oxford
22. Whipple Museum of the History of Science, University of Cambridge
23. Wikimedia Commons
24. © National Portrait Gallery, London
25. © Dan Winters/Courtesy of the Computer History Museum
26. Science Museum Library/Science & Society Picture Library
27. © National Portrait Gallery, London

28   Oxford, Bodleian Library, Dep. Lovelace Byron 117, fols 49v–48r
29   Oxford, Bodleian Library, Dep. Lovelace Byron 77, fols 218v–217r
30   © Marcin Wichary/Courtesy of the Computer History Museum
31   Oxford, Bodleian Library, Dep. Lovelace Byron 170, fols 100v–101r
32   UCL Library Services, Special Collections, MS ADD 7
33   Oxford, Bodleian Library, Dep. Lovelace Byron 170, fols 12v–13r
34   Private collection
35   Oxford, Bodleian Library, Dep. Lovelace Byron 170, fols 91v and 92r
36   Oxford, Bodleian Library, Dep. Lovelace Byron 170, fol. 11r
37   Oxford, Bodleian Library, Dep. Lovelace Byron 170, fols 5v–6r
38   Oxford, Bodleian Library, Dep. Lovelace Byron 170, fol. 354v
39   Oxford, Bodleian Library, 1811 e.35
40   By permission of the Royal Irish Academy © RIA
41   Oxford, Bodleian Library, Dep. Lovelace Byron 170, fol. 128r
42   Oxford, Bodleian Library, Dep. Lovelace Byron 170, fol. 19v
43   Oxford, Bodleian Library, Dep. Lovelace Byron 170, fol. 143r
44   © Museum of the History of Science, University of Oxford, MS. Buxton 7
45   Science Museum Library/Science & Society Picture Library
46   © Sydney Padua, *The Thrilling Adventures of Lovelace and Babbage*, Particular Books, 2015
47   Magdalen College Libraries and Archives, Daubeny 90.A.11, reproduced by permission of The President and Fellows of Magdalen College, Oxford, Daubeny
48   Oxford, Bodleian Library, Dep. Lovelace Byron 170, fol. 179r
49   Oxford, Bodleian Library, Dep. Lovelace Byron 168, fols 45v and 46r
50   London, British Library, MS. Add 37192, fols 362v–363.
     © British Library Board. All Rights Reserved/Bridgeman Images
51   © Victoria and Albert Museum, London
52   Ostpreußisches Landesmuseum, Lüneburg/Dr Andras Vieth
53   Oxford, Bodleian Library, Dep. Lovelace Byron 170, fols 176v–177r
54   Wikimedia/Creative Commons
55   Oxford, Bodleian Library, Dep. Lovelace Byron 44, fol. 210r
56   Oxford, Bodleian Library, (Vet.) 177 a.1
57   Oxford, Bodleian Library, Mary Somerville Papers, MS. Dep. c. 367, Folder MSBY-9, fol. 255r
58   Private collection
59   Park Dale/Alamy Stock Photo

INDEX

Adams, George 31
Albert of Saxe-Coburg and Gotha, Prince Consort 99

Babbage, Charles 6, 31, 36–45, 49, 62, 73, 75–9, 84–9, 92–9, 100–103
   Analytical Engine 62, 72–87, 102
   Difference Engine 36, 37–8, 41, 44, 44–5
   Reflections on the Decline of Science in England, and on Some of Its Causes 37
   The Economy of Machinery and Manufactures 37
Bonnycastle, John 24
British Association for the Advancement of Science 3
Brunel, Isambard Kingdom 44
Byron, Ada *see* Lovelace
Byron, Anne Isabella, Lady, *née* Milbanke 1–6, 9–10, 13, 17, 23–7, 34, 42–5, 59, 62, 102–4
Byron, George Gordon, 6th Baron 1, 5, 9, 17, 103
   Childe Harold's Pilgrimage 1
   Don Juan 2

Cauchy, Augustin-Louis 61
Cayley, George 20–21
Chambers, Robert 99
   Vestiges of the Natural History of Creation 99, 102
Claudet, Antoine 40, 50
Clement, Joseph 41
Coleridge, Samuel Taylor 6, 96

Darwin, Charles 44, 99
De Morgan, Augustus 6, 7, 45, 49–70, 102–3
   Elements of Algebra 63, 64
   Laws 49
   Principle of Permanence 63–6
De Morgan, Sophia, *née* Frend 45, 49
Dickens, Charles 44

Euler, Leonhard 89–93

Faraday, Michael 86
Frend, William 2, 6, 24–5, 27, 30, 45
Fry, Elizabeth 2

Gasparin, Agénor de 100
Girton College, Cambridge 7

Great Exhibition of 1851 3, 4, 87, 98, 99
Greig, Woronzow 31, 34, 101

Hamilton, William Rowan 64, 66
   quaternions 66
Haydon, Benjamin Robert 6
Herschel, Caroline 25, 31
Herschel, John 37, 100
Herschel, William 25
Jacquard, Joseph Marie 73, 82

King, Dr William 6, 23, 27, 30–31, 43, 44, 93
King, William *see* Lovelace

Lamb, Lady Caroline 2
Lamont, Miss (governess) 9–11, 17
Lardner, Dionysius 44–5
Leonardo da Vinci 17
Lovelace, Augusta Ada, Countess of
   born 1
   taught by Miss Lamont 9–10
   married William King 31
   children Byron, Anabella and Ralph 32, 62
   tutored by De Morgan 49–70
   paper published 77–86
   death 103
   Kirkby Mallory memorial 104
Lovelace, William King, 1st Earl of 31–4, 77, 99
   'Method of Growing Beans and Cabbages on the Same Ground' 99–100
   'On Climate in Connection with Husbandry' 100

Martineau, Harriet 5, 34
   England and Her Soldiers (with Nightingale) 5
Marx, Karl 37
mathematics
   Bernoulli numbers 70–71, 77–81, 83, 85–6
   complex numbers 65–7
   divergent series 69–70
   Euclidian geometry 5, 16, 27, 30
   Königsberg Bridge Problem 89, 92, 93
   logarithms 58
   magic square 92, 94
   method of finite differences 37–41, 46–7
   peg solitaire 95
   Pythagoras' theorem 27, 30, 92–3
   quaternions 66
   rule of three 13–16
   Tit-tat-to 94
Mechanics' Magazine 20
Menabrea, Luigi 77–9
Merian, Matthias 89
Milbanke, Sir Ralph 1
Mill, John Stuart 2
Montgolfier, Joseph-Michel and Jacques-Étienne 17
Mote, William Henry 3

Newton, William John 3
Nightingale, Florence 4, 5
   England and Her Soldiers (with Martineau) 5

Orsay, Count Alfred d' 9

Padua, Sydney 76
Pestalozzi, Johann Heinrich 2, 10–11, 23, 34
Phillips, Henry Wyndham 103
Poinsot, Louis 93
Priestley, Joseph 2
Purser, Sarah 64

Quetelet, Adolphe 100

Redgrave, Richard 10
Royal Astronomical Society 31

Royal Statistical Society 99

Sargent, Frederick 34
Somerville, Mary, *née* Fairfax 7, 30–34, 99, 102–3
    translation of *Mécanique Céleste* 31
Somerville College, Oxford 7
Sturge, William 6

Toothill, Geoff 79

Trevithick, Richard 3
    'Puffing Devil' 3
Turing, Alan 73, 82
    'Lady Lovelace's objection' 82

University College, London 49

Westall, Richard 1
Wheatstone, Charles 45, 77
Whewell, William 99
World Anti-Slavery Convention 6

Diagram for the computation by the Engine of

Number of Operation.	Nature of Operation.	Variables acted upon.	Variables receiving results.	Indication of change in the value on any Variable.	Statement of Results.	Data. $^1V_1$ ○ 0 0 1 [1]	$^1V_2$ ○ 0 0 2 [2]	$^1V_3$ ○ 0 0 4 [n]	$^0V_4$ ○ 0 0 0 ☐	$^0V_5$ ○ 0 0 0 ☐	$^0V_6$ ○
1	×	$^1V_2 \times {}^1V_3$	$^1V_4, {}^1V_5, {}^1V_6$	$\left\{\begin{array}{l}{}^1V_2={}^1V_2\\{}^1V_3={}^1V_3\end{array}\right\}$	$= 2n$	...	2	n	$2n$	$2n$	$2n$
2	−	$^1V_4 - {}^1V_1$	$^2V_4$	$\left\{\begin{array}{l}{}^1V_4={}^2V_4\\{}^1V_1={}^1V_1\end{array}\right\}$	$= 2n-1$	1	...	...	$2n-1$	...	...
3	+	$^1V_5 + {}^1V_1$	$^2V_5$	$\left\{\begin{array}{l}{}^1V_5={}^2V_5\\{}^1V_1={}^1V_1\end{array}\right\}$	$= 2n+1$	1	...	...	...	$2n+1$	...
4	÷	$^2V_5 \div {}^2V_4$	$^1V_{11}$	$\left\{\begin{array}{l}{}^2V_5={}^0V_5\\{}^2V_4={}^0V_4\end{array}\right\}$	$= \dfrac{2n-1}{2n+1}$	...	...	...	0	0	...
5	÷	$^1V_{11} \div {}^1V_2$	$^2V_{11}$	$\left\{\begin{array}{l}{}^1V_{11}={}^2V_{11}\\{}^1V_2={}^1V_2\end{array}\right\}$	$= \dfrac{1}{2} \cdot \dfrac{2n-1}{2n+1}$	...	2	...	...	...	...
6	−	$^0V_{13} - {}^2V_{11}$	$^1V_{13}$	$\left\{\begin{array}{l}{}^2V_{11}={}^0V_{11}\\{}^0V_{13}={}^1V_{13}\end{array}\right\}$	$= -\dfrac{1}{2} \cdot \dfrac{2n-1}{2n+1} = A_0$	...	...	...	...	...	...
7	−	$^1V_3 - {}^1V_1$	$^1V_{10}$	$\left\{\begin{array}{l}{}^1V_3={}^1V_3\\{}^1V_1={}^1V_1\end{array}\right\}$	$= n-1 (= 3)$	1	...	n	...	...	...
8	+	$^1V_2 + {}^0V_7$	$^1V_7$	$\left\{\begin{array}{l}{}^1V_2={}^1V_2\\{}^0V_7={}^1V_7\end{array}\right\}$	$= 2+0 = 2$	...	2	...	...	...	...
9	÷	$^1V_6 \div {}^1V_7$	$^3V_{11}$	$\left\{\begin{array}{l}{}^1V_6={}^1V_6\\{}^0V_{11}={}^3V_{11}\end{array}\right\}$	$= \dfrac{2n}{2} = A_1$	...	...	...	...	...	2
10	×	$^1V_{21} \times {}^3V_{11}$	$^1V_{12}$	$\left\{\begin{array}{l}{}^1V_{21}={}^1V_{21}\\{}^3V_{11}={}^3V_{11}\end{array}\right\}$	$= B_1 \cdot \dfrac{2n}{2} = B_1 A_1$	...	...	...	...	...	...
11	+	$^1V_{12} + {}^1V_{13}$	$^2V_{13}$	$\left\{\begin{array}{l}{}^1V_{12}={}^0V_{13}\\{}^1V_{13}={}^2V_{13}\end{array}\right\}$	$= -\dfrac{1}{2} \cdot \dfrac{2n-1}{2n+1} + B_1 \cdot \dfrac{2n}{2}$	...	...	...	...	...	...
12	−	$^1V_{10} - {}^1V_1$	$^2V_{10}$	$\left\{\begin{array}{l}{}^1V_{10}={}^2V_{10}\\{}^1V_1={}^1V_1\end{array}\right\}$	$= n-2 (= 2)$	1	...	...	...	...	...
13	−	$^1V_6 - {}^1V_1$	$^2V_6$	$\left\{\begin{array}{l}{}^1V_6={}^2V_6\\{}^1V_1={}^1V_1\end{array}\right\}$	$= 2n-1$	1	...	...	...	...	$2n$
14	+	$^1V_1 + {}^1V_7$	$^2V_7$	$\left\{\begin{array}{l}{}^1V_1={}^1V_1\\{}^1V_7={}^2V_7\end{array}\right\}$	$= 2+1 = 3$	1	...	...	...	...	...
15	÷	$^2V_6 \div {}^2V_7$	$^1V_8$	$\left\{\begin{array}{l}{}^2V_6={}^2V_6\\{}^2V_7={}^2V_7\end{array}\right\}$	$= \dfrac{2n-1}{3}$	...	...	...	...	...	$2n$
16	×	$^1V_8 \times {}^3V_{11}$	$^4V_{11}$	$\left\{\begin{array}{l}{}^1V_8={}^0V_8\\{}^3V_{11}={}^4V_{11}\end{array}\right\}$	$= \dfrac{2n}{2} \cdot \dfrac{2n-1}{3}$	...	...	...	...	...	...
17	−	$^2V_6 - {}^1V_1$	$^3V_6$	$\left\{\begin{array}{l}{}^2V_6={}^3V_6\\{}^1V_1={}^1V_1\end{array}\right\}$	$= 2n-2$	1	...	...	...	...	$2n$
18	+	$^1V_1 + {}^2V_7$	$^3V_7$	$\left\{\begin{array}{l}{}^2V_7={}^3V_7\\{}^1V_1={}^1V_1\end{array}\right\}$	$= 3+1 = 4$	1	...	...	...	...	...
19	÷	$^3V_6 \div {}^3V_7$	$^1V_9$	$\left\{\begin{array}{l}{}^3V_6={}^3V_6\\{}^3V_7={}^3V_7\end{array}\right\}$	$= \dfrac{2n-2}{4}$	...	...	...	...	...	$2n$
20	×	$^1V_9 \times {}^4V_{11}$	$^5V_{11}$	$\left\{\begin{array}{l}{}^1V_9={}^0V_9\\{}^4V_{11}={}^5V_{11}\end{array}\right\}$	$= \dfrac{2n}{2} \cdot \dfrac{2n-1}{3} \cdot \dfrac{2n-2}{4} = A_3$	...	...	...	...	...	...
21	×	$^1V_{22} \times {}^5V_{11}$	$^0V_{12}$	$\left\{\begin{array}{l}{}^1V_{22}={}^1V_{22}\\{}^0V_{12}={}^2V_{12}\end{array}\right\}$	$= B_3 \cdot \dfrac{2n}{2} \cdot \dfrac{2n-1}{3} \cdot \dfrac{2n-2}{3} = B_3 A_3$	...	...	...	...	...	...
22	+	$^2V_{12} + {}^2V_{13}$	$^3V_{13}$	$\left\{\begin{array}{l}{}^2V_{12}={}^0V_{12}\\{}^2V_{13}={}^3V_{13}\end{array}\right\}$	$= A_0 + B_1 A_1 + B_3 A_3$	...	...	...	...	...	...
23	−	$^2V_{10} - {}^1V_1$	$^3V_{10}$	$\left\{\begin{array}{l}{}^2V_{10}={}^3V_{10}\\{}^1V_1={}^1V_1\end{array}\right\}$	$= n-3 (= 1)$	1	...	...	...	...	...

Here follows a repetiti

| 24 | + | $^4V_{13} + {}^0V_{24}$ | $^1V_{24}$ | $\left\{\begin{array}{l}{}^4V_{13}={}^0V_{13}\\{}^0V_{24}={}^1V_{24}\end{array}\right\}$ | $= B_7$ | ... | ... | ... | ... | ... | ... |
| 25 | + | $^1V_1 + {}^1V_3$ | $^1V_3$ | $\left\{\begin{array}{l}{}^1V_1={}^1V_1\\{}^1V_3={}^1V_3\\{}^5V_6={}^0V_6\\{}^5V_7={}^0V_7\end{array}\right\}$ | $= n+1 = 4+1 = 5$ .... by a Variable-card. by a Variable card. | 1 | ... | $n+1$ | ... | ... | 0 |